New Age Analytics

New Age Analytics

Alvin Albuero De Luna

www.arclerpress.com

New Age Analytics

Alvin Albuero De Luna

Arcler Press

224 Shoreacres Road

Burlington, ON L7L 2H2

Canada

www.arclerpress.com

Email: orders@arclereducation.com

© 2021 Arcler Press

ISBN: 978-1-77407-711-5 (Hardcover)

Arcler Press publishes wide variety of books and eBooks. For more information about Arcler Press and its products, visit our website at www.arclerpress.com

ABOUT THE AUTHOR

Alvin Albuero De Luna is an instructor at a Premier University in the Province of Laguna, Philippines - the Laguna State Polytechnic University (LSPU). He finished his Bachelor's degree in Information Technology at STI College and took his Master of Science in Information Technology at LSPU. He isnhandling Programming Languages, Cyber Security, Discrete Mathematics, CAD, and other Computer related courses under the College of Computer Studies

TABLE OF CONTENTS

GLOSSARY

A

Abstraction – the process of removing physical, spatial, or temporal details.

Algorithm – a process or set of rules to be followed in calculations or other problem-solving operations, especially by a computer.

Annotations – the action of annotating a text or diagram.

Anomaly – something that deviates from what is standard, normal, or expected.

Audiotape – magnetic tape on which sound can be recorded.

Automate – a process or facility, to be operated by largely automatic equipment.

B

Bias – cause to feel or show inclination or prejudice for or against someone or something.

Big Data – a field that treats ways to analyze, systematically extract information from or otherwise deal with data sets that are too large or complex to be dealt with by traditional data-processing application software.

Biography – a detailed description of a person's life.

C

Compatible – (of two things) able to exist or occur together without problems or conflict.

Concurrently – at the same time; simultaneously.

Curtailing – to make less by or as if by cutting off or away from some part.

Customer Relationship Management – Customer relationship management is an approach to manage a company's interaction with current and potential customers. It uses DA about customers' history with a company to improve business relationships with customers, specifically focusing on customer retention and ultimately driving sales growth.

D

Data Mining – the practice of examining large pre-existing databases in order to generate new information.

Demographics – statistical data relating to the population and particular groups within it.

Digital Assistance – is a computer program designed to assist a user by answering questions.

Disruptive – causing or tending to cause disruption.

E

eCommerce – is the activity of electronically buying or selling of products on online services or over the Internet.

Encumbered – restrict or impede (someone or something) in such a way that free action or movement is difficult.

Entrepreneurs – an individual who creates a new business, bearing most of the risks and enjoying most of the rewards.

Expedition – promptness or speed in doing something.

F

Fallibility – the tendency to make mistakes or be wrong.

Forecasting – the process of making predictions of the future based on past and present data and most commonly by analysis of trends.

G

Grids – a framework of spaced bars that are parallel to or cross each other.

I

Inferences – a conclusion reached on the basis of evidence and reasoning.

Integration – the action or process of integrating.

J

Jargon – special words or expressions used by a profession or group that are difficult for others to understand.

L

Label – Label is the thing is to be predicted with the help of models.

M

Metadata – a set of data that describes and gives information about other data.

Models – A model is a relationship between features and the label.

O

Optimization – the action of making the best or most effective use of a situation or resource.

Overcomplicated – make (something) more complicated than necessary.

P

Perpetuated – make (something) continue indefinitely

Platform – A platform is used for automating and accelerating the delivery lifecycle of predictive applications capable of processing big data using machine learning or related techniques.

Profiling – the recording and analysis of a person's psychological and behavioral characteristics, so as to assess or predict their capabilities in a certain sphere or to assist in identifying categories of people.

Q

Questionnaire – a set of printed or written questions with a choice of answers, devised for the purposes of a survey or statistical study.

R

Recognition – identification of someone or something or person from previous encounters or knowledge.

Regression Trees – A Regression tree may be considered as a variant of decision trees, designed to approximate real-valued functions, instead of being used for classification methods.

Remuneration – money paid for work or a service.

Repetitive jobs – involves actions or elements that are repeated many times and is therefore boring.

Revenue – income, especially when of an organization and of a substantial nature

S

Scalable – able to be scaled or climbed.

Semantic Ambiguity – It happens when a word, phrase or sentence, taken out of context, has more than one interpretation.

Semantic Computing – Semantic computing is a field of computing that combines elements of semantic analysis, natural language processing, data mining, and related fields.

Simulate – an approximate imitation of the operation of a process or system; that represents its operation over time.

Social Media Analytics – Social media analytics is the process of gathering and analyzing data from social networks such as Facebook, Instagram, and Twitter. It is commonly used by marketers to track online conversations about products and companies.

Social Media Intelligence – Social media intelligence refers to the collective tools and solutions that allow organizations to and conversations, respond to social signals and synthesize social data points into meaningful trends and analysis based on the user's needs.

Sophisticated – having, revealing, or involving a great deal of worldly experience and knowledge of fashion and culture.

Succinct – especially of something written or spoken) briefly and clearly expressed.

Supply Chain – the sequence of processes involved in the production and distribution of a commodity.

U

Unobtrusive research – a method of data collection used primarily in the social sciences.

LIST OF FIGURES

LIST OF TABLES

LIST OF ABBREVIATIONS

AI	artificial intelligence
ALS	amyotrophic lateral sclerosis
ANN	artificial neural network
B2B	business to business
BA	business analytics
BI	business intelligence
CARET	classification and regression training
CRM	customer relationship management
DA	data analysis
DBMSs	database management systems
DNA	deoxyribonucleic acid
DSMSs	data stream management systems
E-CRM	electronic customer relationship management
EDA	exploratory data analysis
GPUs	graphics processing units
HPC	high performance computing
IB	business intelligence
INFORMS	Institute of Operations Research and Management Sciences
IoT	internet of things
IT	information technology
KPIs	key performance indicators
LDA	latent dirichlet allocation
LDW	logical data warehouse
MALLET	machine learning for language toolkit
MATLAB	matrix laboratory
ML	machine learning
MSHS	Master of Science and Health Science

NLTK	natural language toolkit
OCR	optical character recognition
PMI	point wise mutual information
PMO	project management office
POS	part of speech
PR	public relation
PwC	pricewaterhouse coopers
RAI	retailers association of India
RBV	resource-based view
ROI	return on investment
SciPy	scientific python
SMA	social media analytics
SOM	self-organizing maps
SPSS	statistical package for the social sciences
TPU	tensor processing unit

PREFACE

The world is progressing at a very rapid pace and the processes are getting aligned in a digital manner. The leaders and entrepreneurs realize that it is extremely important to know about the situations in detail so that they can design the processes and prepare themselves for the future with accuracy and certainty.

To know about the situations with accuracy and precision, the professionals and the leaders need to analyze the situations and figures related to them in detail, so that they can reach a conclusion about the strategies they need to take to register good profits and have good returns from the investments they make.

Thus, analytics holds a great place in the modern business and theoretical world, which makes it an important field of study for the upcoming digital enthusiasts and researchers. Analytics provides the people with a look into the future and makes them able to take prompt and informed decisions, which reduce their risks related to investment.

The digital analytics has advanced as a field with a rapid pace and has become a popular subject among the youth, that believes in investing time and money in a smart way. Youth find it helpful to analyze the market and the related things so that they are fully aware of the trends in the market and the inclination of the customers.

The new age analytics refers to the methods of analytics used that go through all the parameters through which a situation or a thing can be analyzed and focuses on all the aspects, so that there is no space that is left behind and no room for misinformation.

This book takes the readers through the various concepts in the field of analytics and makes them aware of the various methods in the new age analytics. The book throws light on the several aspects that are related to the new age analytics and tries to make the readers about them in the most detailed manner.

The first chapter of the book starts off by introducing the meaning of analytics to the readers and making them aware of the various techniques that are used in analytics. The second chapter relates analytics to the field of digital marketing so that the readers can see analytics in light of marketing and promoting various goods and services for the consumers. The book focuses on the concepts related to artificial intelligence and takes the readers through the various ways in which artificial intelligence is governed by analytics and how the new age analytics develops the much-awaited and anticipated field of artificial intelligence.

The fourth chapter informs the readers about the use of tools, such as that of machine learning, to analyze the various aspects of a business so that the parts requiring a huge amount of data can be analyzed with precision and in stipulated time.

The fifth chapter takes the readers through the application of analytics to the field of social media, which has a great deal of application in the digital world. Then the book talks about the application of IOT in business analytics (BA) and informs the readers about the importance of the IOT tools in this field.

The readers are then told about the process of customer acquisition in the retail market and how such analytics is used to analyze the customer base, in the seventh chapter. In the eighth chapter, the book talks about the future of analytics, highlighting the reasons that make it a field to look forward to in the future.

This book tries to make the readers well-informed on the subject of new age analytics and makes efforts towards upgrading the knowledge of the readers in this regard. I hope that the readers find the book enlightening and fulfilling in terms of knowledge and that this book can truly serve its purpose of spreading information and knowledge.

Analysis and Its Techniques

CONTENTS

The chapter of analysis and its techniques explains about the basic significance of the analysis. This chapter also mentions the various types of analysis in research, which is consisting of explanatory data analysis, descriptive data analysis, casual data analysis, predictive data analysis, inferential data analysis, decision trees, mechanistic data analysis, and evolutionary programming.

This chapter emphasizes the various phases of the analysis such as data requirement gathering, data collection, data cleaning, data analysis, data interpretation, and data visualization. This chapter explains the various methods to perform data analysis, and also provide brief about qualitative data, and quantitative data, what are the data strategies to conceptualize the data analysis.

1.1. INTRODUCTION

A process of cleaning, transforming, and modeling data in order to discover the information which is useful for business decision – making is known as data analysis (DA). The main aim of the DA is to extract useful information from data and take the decision on the basis of the DA.

Figure 1.1. Basic significance of the Data analysis.

Source: Image by The Blue Diamond Gallery.

In our day to day life, whenever there is any decision taken by thinking about what happened last time or what is going to happen by choosing that specific decision. This is nothing but an analysis that is done of our past or future and making decisions based on it. For that particular matter, memories are gathered from our past or dreams of our future. So, that is just

DA and nothing. The same thing is done by the analyst for the purpose of a business that is referred to as DA.The tools used in DA make it easier for the users to process as well as manipulate data, analyze the relationships as well as correlations in between the data sets and it also, helps in identifying the patterns and trends in order to interpretThe DA is defined as the process of applying the statistical or logical techniques in a systematic manner so as to describe and illustrate, condense, and recap and also, evaluate the data. According to Shamoo and Resnik, (2003), there are a number of analytic procedures which provide a way of drawing the inductive inferences from the data and distinguishing the signal (that is the phenomenon of interest) from the noise (that is the statistical fluctuations) which are present in the data.

In the qualitative research, the DA can include the statistical procedures, many times analysis becomes an ongoing iterative process in which the data is continuously collected as well as analyzed almost at the same time. Generally, the researchers analyze the patterns in the observations through the entire phase of data collection (Savenye and Robinson, 2004).

The form of analysis can be determined by the particular qualitative approach which is taken at the time of analysis (field study, ethnography content analysis, oral history, biography, unobtrusive research) and the form of data on which the analysis is to be done (field notes, documents, audiotape, videotape).

In order to ensure data integrity, an essential component is the accurate and appropriate analysis of research findings. The improper statistical analyses distort the scientific findings and so, mislead the casual readers (Shepard, 2002), and also, can influence the public perception of research in a negative manner.

1.1.1. Definitions of Analysis

> According to the Cambridge Dictionary of Philosophy, 2nd ed., 1999, ed. Robert Audi:
>
> "The process of breakingup a concept, proposition, linguistic complex, or fact into its simple or ultimate constituents."

According to the Concise Oxford Dictionary, 1976, ed. J. B. Sykes:

1. Resolution into simpler elements by analyzing; statement of the result of this; and

2. (Math.) Use of algebra and calculus in problem-solving.

According to the Dictionary of Philosophy and Psychology, 1925, ed. James Mark Baldwin:

"The isolation of what is more elementary from what is more complex by whatever method."

According to the A Kant Dictionary, 1995, by Howard Caygill:

Two senses were combined by Kant of analysis in his work. From those two, one was derived from Greek geometry, the other was derived from modern physics and chemistry. Both the senses remain close to the original Greek sense of analysis which is 'loosening up' or 'releasing,' but each proceeds in different ways.

According to the Oxford Dictionary of Philosophy, 1996, by Simon Blackburn:

"The process of breaking a concept down in to more simple parts, so that its logical structure is displayed."

As per the view of Routledge Encyclopedia of Philosophy, 1998, by Thomas Baldwin:

"Philosophical analysis is defined as a method of inquiry in which one seeks to assess complex systems of thought by 'analyzing' them into simpler elements whose relationships are brought into focus."

And by Robert Hanna:

"The conceptual analysis holds that concepts (which is the general meanings of linguistic predicates), are considered to be the fundamental objects of philosophical inquiry, and those insights into conceptual contents are expressed in necessary 'conceptual truths' (analytic propositions)."

1.2. EIGHT TYPES OF ANALYSIS IN RESEARCH

DA is said to be useful in drawing certain conclusion about the variables which are there in the research. However, the approach to analysis depends on the research which is being carried out. If data analytics is not used, then determining the relationship between the variables becomes difficult which eventually leads to a meaningful conclusion. Therefore, DA is said to be an important tool so as to achieve a particular conclusion.

There are different types of analysis if the research is considered, and those are mentioned below:

1.2.1. Exploratory Data Analysis (EDA)

Among a number of DA types, there is one named exploratory DA, which is used in order to analyze the data as well as relationships that are established and were previously not known. Specifically, they are used in order to discover new connections and to define future studies or answer the questions which pertain to future studies.

By performing the exploratory analysis, the answers are provided, and those answers are not definitive in nature, but they provide a little information into what is coming. Using visual methods, the approach to analyzing the data sets is the commonly used technique for EDA. John Tukey promotes the exploratory DA and was defined in the year 1961.

There are graphical techniques for representation, and these are used primarily in exploratory DA. Some of the most used graphical techniques are histogram, stem, and leaf plot, Pareto chart, scatter plot, box plot and many more. There are some drawbacks also in the exploratory analysis and one of them is that it cannot be used for generalizing or predicting the precisely about the upcoming events.

The data provides the correlation in which the causation does not imply. The exploratory DA can be applied while studying the census along with the convenience of the sample data sets.

In the EDA system, software, and machine-aided have become very usual and some of them are Data Applied, Python, Ggobi, KNIME, JMP, etc.

1.2.2. Descriptive Data Analysis

Among all the other methods, the method of descriptive DA requires the least amount of effort. Furthermore, it describes the main features of the collection of data in a quantitative manner. Usually, this is the initial kind of DA, which is performed on the data set which is available.

Usually, the descriptive DA is applied to the volumes of data like the census data. There are different steps involved in the descriptive DA for description as well as interpretation. There are two methods of statistical descriptive analysis which are univariate as well as bivariate. Both are considered to be the types of analysis in research.

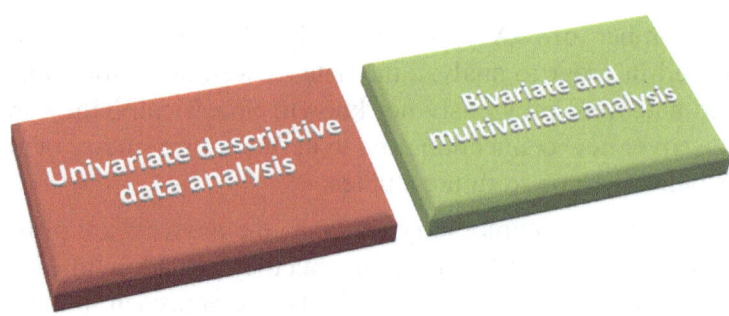

Figure 1.2. Types of descriptive data analysis.

- **Univariate descriptive data analysis**: This is the type of analysis that involves the distribution of a single variable and that is referred to as univariate analysis.

- **Bivariate and multivariate analysis:** The bivariate and multivariate analysis can be defined as the data analysis which involves a description of the distribution of more than one variable. In such cases, the descriptive statistics may be used for describing the relationship between the pair of variables.

1.2.3. Causal Data Analysis

Causal DA is also referred to as explanatory DA. The word causal determines the cause and effect relationship that is between the variables. Primarily, the analysis is carried out in order to see in the case, if one variable would change, then what would happen to another variable.

Usually, randomized studies are required in the application of causal studies but there are also approaches for concluding the causation even as well as non-randomized studies.

It is considered to be very complex and the researcher cannot be certain that other variables, which influence the causal relationship, are constant especially when the research is dealing with the attitudes of customers in the business.

In most cases, the researcher is required to consider the psychological impacts that even the respondent cannot be aware of at any instance and these parameters, which are not considered, impact the data which is analyzed and can affect the conclusions.

1.2.4. Predictive Data Analysis

Predictive DA is the type of DA that involves using the methods which analyze the current trends along with the historical facts in order to arrive at a conclusion that makes predictions about the future trends of the upcoming future events.

The prediction, as well as the success of the model, depend on choosing and measuring the right variables. It is quite difficult to predict future trends and so, it requires technical expertise in the subject. Machine learning (ML) is referred to as a modern tool that is used to do the interactive analysis for getting better results. Prediction analysis is used to predict the rising as well as changing trends in a number of various industries.There are many applications of predictive DA such as Analytical customer relationship management (CRM), collection analytics, clinical decision support systems, portfolio management, and fraud detection. There is one more important application of the predictive DA and that is forecasting the future financial trends. There are software which is used for Predictive Analysis such as Apache Mahout, GNU Octave, OpenNN, MATLAB, and many more.

1.2.5. Inferential Data Analysis

Inferential DA is one of those types of analysis in research that tests theories of different subjects on the basis of the sample taken from the group of subjects. Under this analysis, a small part of the population is studied, and the results are applied to the bigger population size. The small chunk of the population is chosen from different backgrounds and tastes so that the results could be applied to a large population.

The small chunk of the population is studied, and conclusions are drawn thereafter. The main goal of the statistical models is to provide inference and a conclusion. The selection of a proper statistical model for the process is very important because the process involves drawing conclusions.

The sole reason for the success of inferential DA is the selection of the proper statistical model used for analysis. The sampling technique and the population determines the results of the inferential analysis. Therefore, it is important to choose a variety of representative subjects for better results.

The DA is applied to the cross-sectional study of time, retrospective data set and observational DA for the research purpose. Along with proper sampling techniques, good tools must also be used for providing proper results of inferential DA.

1.2.6. Decision Trees

Decision trees are one of those types of analysis which are classified as a modern classification algorithm in the data mining. It is considered to be a very popular type of analysis in research which further requires ML. Usually, it is represented as a tree-shaped diagram of a figure, which provides information about the regression models or classification.

The decision tree can be subdivided into the smaller database, which has similar values. The determination of the branches is done on how the tree is built where does one goes with the current choices and where would those choices lead to next.

The primary advantage of a decision tree is the domain knowledge, which is not considered to be an important requirement for the purpose of analysis. In addition to this, the classification of the decision tree is a very simple as well as fast process which takes less time in comparison to other DA techniques.

1.2.7. Mechanistic Data Analysis

The method of mechanistic DA is exactly opposite to the descriptive DA, which requires the least amount of effort, and the mechanistic DA requires a maximum amount of effort. The main idea behind the mechanistic DA is to understand the nature of some exact changes in the variables which affect some other variables.

The mechanistic DA is considered to be exceptionally difficult in predicting the except when the situations are much simpler. This analysis is used by the physical and engineering science and this is used in the case of the deterministic set of equations. The application of this kind of analysis is randomized trial data set.

1.2.8. Evolutionary Programming

The evolutionary programming combines the different types of analysis in research by using the evolutionary algorithms in order to form the meaningful data and so, it is considered to be a very common concept in the data mining.

The genetic algorithms as well as evolutionary algorithms are the most popular programs of the revolutionary programming. In the case of independent techniques, these are an accident because of the fact that they have the ability to search as well as explore the large spaces so as to discover good solutions.

1.3. PHASES IN ANALYSIS

Figure 1.3. Different phases in the process of analysis.

There are six phases in the process of data analysis:

- Data Requirement Gathering;
- Data Collection;
- Data Cleaning;
- Data Analysis;
- Data Interpretation; and
- Data Visualization.

1.3.1. Data Requirement Gathering

First and foremost, it is important to think about why anyone wants to do this DA. Everything that one needs to find out is the purpose or aim of performing the analysis. It is important to decide which type of DA, they want to perform on the data. In such a phase, it is required to decide what should be analyzed and how to measure it, it should be understood why you are investigating and what measures are required to do this analysis.

1.3.2. Data Collection

After gathering the requirement, the analyst will get an idea about what things they have to measure and what should be the findings be like. Then there comes the time to collect the data on the basis of requirements. Once, the collection of the data is done, it is important to remember that the data which is collected, must be processed or organized for performing analysis. As the data is collected from various sources, it is important to keep a log with a collection data and source of the data.

1.3.3. Data Cleaning

Now the data that is collected may not be useful or it may be irrelevant to the aim of Analysis. Hence, this data should be cleaned. The data that is gathered may contain some errors, duplicate records or white spaces. So, this data should be error-free and cleaned. Before the process of analysis, this phase must be done because based on data cleaning, the output of Analysis will be closer to the outcome that is expected.

1.3.4. DA

The data is ready for analysis once it is gathered, cleaned, and processed. During the manipulation of data, an individual may find that they have the exact information that is required, or you might need to gather more information. During this period, one can make use of DA tools and software. This which will help an individual to understand, interpret, and derive conclusions that are based on the requirements.

1.3.5. Data Interpretation

It is a final time to interpret the outcomes. This phase comes after analyzing the data. An individual can choose the method in order to express or communicate the DA. They can either express or communicate it by simply using the words or maybe it can be expressed by the help of table or chart. Later on, the outcomes of the process of DA are used to decide the best course of action.

1.3.6. Data Visualization

Data visualization is very common method that is used in the day to day life of an individual. Often, this appears in the form of charts and graphs. In other terms, data is shown on graphs so that it will become easy for the human brain to understand and process this data.

Often, this process of data visualization is used to discover facts and trends that are unknown. An individual can find a way to find out important information by examining relationships and comparing all the datasets.

1.4. DA METHODS

DA methods focus on strategic concepts for every single step of DA. From the first step of taking raw data to mine the data for insights that are found to be relevant to the primary goals of the business, and then drilling down into this information so as to transform metrics, facts, and figures into initiatives that benefit improvement.

There are several methods that are used for DA. This is largely based on two main areas:

- Quantitative DA methods; and
- Data analysis methods in qualitative research.

1. **Collaborate Your Needs:** Before an individual starts to analyze the data or drill down into any analysis techniques, it is very important for an individual to sit down collaboratively with all important stakeholders that are inside the organization. They take the decision on the main campaign or strategic goals and attain a fundamental understanding of the types of insights that will provide benefit to the progress or level of vision that is required to evolve the organization.

2. **Establish Your Questions:** Once the main objectives are outlined, an individual should consider all the questions that require answering. This is done in order to help in attaining the mission. This is considered as one of the most important data analytics techniques. It is because this technique will shape the very foundations of success.

3. **Harvest Your Data:** After providing the real direction to the data analytics methodology and finding out which questions require answering to extract optimum value from the information that is available to the organization, an individual should decide on the most valuable sources of data and begin to gather the insights. Among all the techniques, this is the most important DA techniques.

4. **Set Your KPIs:** Once the sources of data are set, one should start collecting the raw data that is considered to offer potential value. In addition to it, clear-cut questions are decided that are required to be answered. Also, there is a requirement to set a host of key performance indicators (KPIs). This will help in tracking, measuring, and shaping the progress in a number of important areas.

KPIs are important to both DA methods in quantitative research as well as DA methods in qualitative research This is considered as one of the important methods of analyzing the data that an individual should not overlook.

5.   **Omit Useless Data:** Having bestowed your DA techniques and methods with true purpose and defined your mission, you should explore the raw data you've collected from all sources and use your KPIs as a reference for chopping out any information you deem to be useless.

The cutting of the informational fat is considered as one of the most important methods of DA. This is because it allows an individual to pay attention to the analytical efforts and extract all the value from the remaining lean information.

6.   **Conduct Statistical Analysis:** Statistical analysis is one of the most important kinds of DA methods. This type of analysis method emphasizes features that include cohort, cluster, regression, factor, and neural networks. In addition to it, this will ultimately provide more logical direction to the DA methodology.

Here is a quick glossary of the statistical analysis terms that are important for the reference:

- **Cluster:** The term cluster is defined as the act of grouping a set of elements in a manner that said elements are more similar (in a particular sense) to each other as compared to the elements that are in the other groups.

- **Cohort:** This is defined as a subset of behavioral analytics that takes insights from a data set that is available (for instance, a web application or CMS) rather than looking at everything as one wider unit. Each element is broken down into groups that are associated with each other.

- **Regression:** A definitive set of statistical processes that are focused on evaluating the relationships among particular variables. This helps in getting a deeper understanding of specific trends or patterns.

- **Factor:** A statistical practice applied in order to describe variability that exists among observed, correlated variables in terms of a potentially lower number of unobserved variables are known as factors. Here, the main purpose is to discover or uncover the independent latent variables.

- **Neural networks:** A neural network is defined as a type of ML which is far too comprehensive to summarize. But it has been observed that this explanation will help in providing a comprehensive picture.

7. **Build a Data Management Roadmap:** At this point, this particular step is optional. This is because the individual has already attained a wealth of insight and created a fairly sound strategy by now. The formation of a data governance roadmap will help the DA methods and techniques to become effective on a more sustainable basis. These roadmaps, if developed properly, are also built so they can be corrected and scaled with the passage of time.

An individual should devote their enough time in developing a roadmap that will help them in storing, managing, and handling the data internally, and to make the analysis techniques all the more fluid as well as functional. This is considered as one of the most powerful kinds of DA methods that are available in the present times.

8. **Integrate Technology:** There are several different ways that are used in order to analyze data. But the integration of the right decision support software and technology is considered as one of the most important aspects of analytical success in a business context.

Robust analysis platforms will not just allow an individual to pull the essential data from the most important sources while working with dynamic KPIs that will provide with actionable insights. It will also present the entire information in a visual, digestible, interactive format from one central, live dashboard.

A data analytics methodology you can count on. If an individual integrates the right technology for the statistical method DA and the core data analytics methodology then he or she will avoid fragmenting the insights, saving their time as well as effort. In addition to it, this will also allow an individual to enjoy the maximum value from the most valuable insights into the business.

9. **Answer Your Questions:** By considering each of the efforts that have been discussed, working with the correct technology, and development of a cohesive internal culture where everybody buys into the different ways to analyze data as well as the power of digital intelligence, one will be able to quickly answer the most burning business questions.

Perhaps, data visualization is the best way that can be used in order to make DA concepts accessible across the organization.

10. **Visualize Your Data:** An online data visualization is considered as powerful as it allows the users across the business to obtain significant insights that help in the evolution of the business. In addition to it, it also covers all the different ways that are used to analyze the data.

The main objective of DA is to make the whole organization more informed and intelligent. With the help of an appropriate platform or dashboard, this process becomes simpler than it is thought over.

1.5. QUALITATIVE DATA

Qualitative data is the data that is represented either in a verbal or narrative form. These kinds of data are gathered with the help of various ways, which include interviews, focus groups, opened-ended questionnaire items, and some other less structured situations.

Thinking of qualitative data in the form of words is a simple way to look at qualitative data. Later on, it will be explored how the transcript can be used as a source of data.

1.5.1. Analyzing Qualitative Data

Qualitative DA works in a little different manner as compared to that of quantitative data. This is mainly because qualitative data is made up of images, observations, words, and even symbols. Deriving absolute meaning from such kind of data is just not possible. Therefore, qualitative data is mostly used for the purpose of exploratory research.

On the other hand, in quantitative research there exist a clear distinction between the stage of data preparation and stage of DA. In the case of qualitative research, analysis often starts as soon as the data is available.

1.5.1.1. Data Preparation and Basic Data Analysis

The process of DA and preparation happen in parallel and it involved the following steps:

1. **Getting Familiar with the Data**: Since it has been observed that most of the qualitative data is just words, the researcher or scientist should begin by reading the data several times in order

to get familiar with it and start to look for basic observations or patterns. In addition to it, this also involves transcribing the data.

2. **Revisiting Research Objectives**: Here, the researcher reconsiders the objective of the research and identifies all the questions that can be answered by the help of data that is collected.

3. **Developing a Framework**: Also called coding or indexing, here the researcher identifies broad concepts, ideas, phrases or behaviors and then a code is assigned to them by the researcher. For instance, gender, coding age, socio-economic status, and even concepts like the positive or negative answer to a particular question. Coding plays a very important role as it helps in structuring as well as labeling the data.

4. **Identifying Patterns and Connections**: Once the data is coded, the researchers can begin to identify the themes, searching for the most common answers to questions, identifying data or patterns that can provide answers to all the research questions, and searching for those areas that can be discovered further.

1.5.2. Qualitative Data Analysis Methods

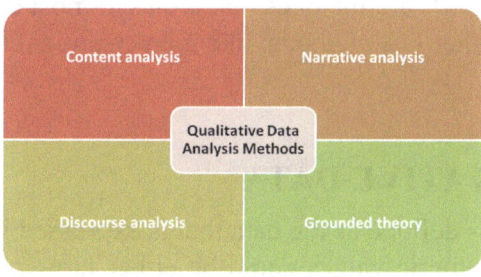

Figure 1.4. Qualitative data analysis methods.

There are various methods available to analyze qualitative data. The DA methods that are used most commonly are:

• **Content analysis**: Among all, this is one of the most common methods that is used to analyze qualitative data. This method is used to analyze documented information in the type of media, texts, or even physical items. When to use this method for analyzing the data is reliant on the research questions. Generally, content analysis is used in order to analyze the answers from interviewees.

- **Narrative analysis**: This method is useful in analyzing the content from different sources, like the observations from the field, interviews of respondents, or surveys. This method emphasizes on using the stories as well as experiences that are shared by the individual to answer the research questions.

- **Discourse analysis**: Discourse analysis is also used to analyze interactions with people just like that of narrative analysis. Though, it emphasizes analyzing the social context in which the interaction between the researcher as well as the respondent takes place. In addition to it, discourse analysis also looks at the day-to-day environment of the respondent and uses that information during the process of DA.

- **Grounded theory**: This refers to the use of qualitative data in order to explain why a certain phenomenon takes place. This is done by examining a variety of cases that are similar in different settings. The data that is collected is used to derive causal explanations. Researchers may change the explanations or make new explanations as they examine more cases until they reach a point where they get the explanation that suits all cases.

These DA methods are the ones that are used most commonly. Though, there are several other DA methods, like conversational analysis, which are also available.

1.6. QUANTITATIVE DATA

Quantitative data is defined as the data that is expressed in numerical terms. In this data, the numeric values could be either large or small. Numerical values may correspond to a particular label or category.

1.6.1. Analyzing Quantitative Data

1.6.1.1. Data Preparation

Data preparation is the initial stage of analyzing the data. The main aim is to convert raw data into something that is meaningful as well as understandable. It involves four steps:

1.6.1.2. Step 1: Data Validation

The objective of data validation is to find out whether the data is collected according to the standards that were set and without any bias. It consists of four steps, that includes:

- **Fraud**, to infer whether every single respondent was interviewed or not;

- **Screening**, to make sure that all the respondents were selected according to the research criteria;

- **Procedure**, to check whether the procedure for the collection of data was followed in a proper manner; and

- **Completeness**, to ensure that the interviewer asked all the questions to the respondent instead of just a few questions that were required.

In order to do this, researchers are required to choose a random sample of completed surveys and validate the data that was collected (it has to be noted that this entire process can be time-consuming for those surveys that have a lot of responses.)

For instance, imagine a survey that included two hundred (200) respondents and it was divided into two cities. The researcher can choose a sample of 20 random respondents from every single city. After this, the researcher can reach out to these respondents by the use of phone or email and check their answers to a certain set of questions that were asked to them.

1.6.1.3. Step 2: Data Editing

Usually, all large data sets involve errors. For instance, respondents may fill some of the fields wrongly or also then can skip some fields accidentally. In order to make sure that there are no such errors in the data set, the researcher should carry out some basic data checks, check for outliers, and edit all the raw research data in order to identify and clear out any data points that may hinder the accuracy of the outcomes.

For instance, an error or mistake could be those fields that were kept empty by respondents. While editing the data, it is very important for an individual to make sure to remove or fill all the fields that are left empty by the respondents.

1.6.1.4. Step 3: Data Coding

In the process of data preparation, this is one of the most important steps. This step indicates to grouping and assigning values to responses that were collected from the survey.

For instance, if a researcher has taken the interview of 1,000 people and now the researcher wants to find the average age of the respondents, the researcher will make age buckets and classify the age of every single respondent as per these codes (for instance, respondents who are between the age group 13–15 years would have their age coded as 0, 16–18 as 1, 18–20 as 2, etc.)

Then during the process of analysis, the researcher can deal with simplified age brackets instead of a vast range of individual ages.

1.6.2. Quantitative Data Analysis Methods

After following all these steps, the data is ready for the process of analysis. The two quantitative DA methods that are mostly used are descriptive statistics and inferential statistics.

1.6.2.1. Descriptive Statistics

Usually, descriptive statistics is the first level of analysis. It is also called as descriptive analysis. It is helpful for the researchers in summarizing the data and finding patterns. Some examples of descriptive statistics that are commonly used are:

- **Mean**: a numerical average of a set of values;
- **Mode**: most common value among a set of values;
- **Median**: midpoint of a set of numerical values;
- **Frequency**: the number of times a value is found;
- **Percentage**: used to express how value or group of respondents inside the data relates to a larger group of respondents; and
- **Range**: the highest as well as the lowest value in a set of values.

Descriptive statistics or descriptive analysis provide absolute numbers. Though, they do not explain the rationale or reasoning that is behind those numbers. It is very important for an individual to think about which one is best suited for the research question and what an individual wants to show, before applying descriptive statistics. For instance, a percentage is considered as a good way to demonstrate the gender distribution of respondents.

Descriptive statistics or descriptive analysis are most useful when the research is restricted to the sample and also, the research does not need to be generalized to a bigger population. For instance, if an individual is comparing the percentage of children those who were vaccinated in two different villages, then descriptive analysis is enough.

Since descriptive analysis or descriptive statistics is mostly used for analyzing a single variable. Often, it is known as univariate analysis.

1.6.3. Data Strategies

There are several numbers of strategies that are used for quantitative and qualitative analyses. Different type of strategies provides an organized approach to the data analysts for working with data. In addition to it, it also enables the data analyst to make a "logical sequence" for the use of the different types of procedures.

1.6.4. Conceptualizing Data Analysis as a Process

What does the term DA actually mean? Does this term refer to one method or many methods? A collection of different kinds of procedures? Is it a procedure? If so, what does that mean? More importantly, can MSHS (Master of Science and Health Science) program staff who are having no background in math or statistics acquire knowledge to identify and apply DA in their work? It should be noted that at least the answer to the last question is must be Yes. This answer is given assuming a minimum investment of time, effort, and practice.

DA can indicate a range of specific procedures as well as methods. Though, before programs can use these specific procedures and methods in an effective manner, it is believed that it is very important to see DA as part of a process.

By this, it is meant that DA includes goals; ideas; relationships; and decision making; in addition to functioning with the actual data itself. In simple terms, DA involves several ways of working with information (data) in order to support the goals, work, and plans of the program or the entire agency.

From this point of view, a DA process is presented that involves the following important components:

- Purpose;
- Questions;

- Data Collection;
- Data Analysis Procedures and Methods;
- Interpretation/Identification of Findings;
- Writing, Reporting, and Dissemination; and
- Evaluation.

From the review of the literature, it has been observed that there are many several numbers of ways of conceptualizing the process of the DA process.

1.   **Data Analysis as a Linear Process**: A strictly linear approach to the process of DA is to work through the components in the sequence, from start to the end. One of the possible advantages of this approach is that this approach structured as well as organized, as all the steps that are involved in the process are organized in a fixed order.

In addition to it, this linear conceptualization of the process may make it easy and simple for the individuals to learn. On the other hand, one of the possible disadvantages is that the step-by-step nature of the decision making may obscure or restrict the power of the analyses. In some other words, the structured nature of the process restricts its efficacy.

2.   **Data Analysis as a Cycle**: A cyclical approach to DA offers much more flexibility to the nature of the decision making and it also involves more and different types of decisions that are to be made. In this cyclic approach, different components that are present in the process can be worked on at different times and it can be worked in different orders if everything comes together at the end.

There are several advantages and disadvantages to this approach. One of the possible advantages of this approach is that program staff are not required to work on every single step in order. The potential that exists for program staff to learn by doing things and to make corrections or improvements to the process before the process is completed.

Thus, the simplest possible answer to the question that was asked above, what is DA, is probably: IT DEPENDS. Instead of deciding to present the process of DA as either linear or cyclical, it has been decided to present both the approaches that are linear and cyclic.

Hopefully, this choice will provide more options and flexibility to the MSHS program staff to make informed decisions, to make use of all those skills that they already have. In addition to it, this choice also helps to grow and develop the ability to use data and its analysis effectively to support the purposes or goals of the program/agency.

1.7. CONSIDERATIONS/ISSUES IN DATA ANALYSIS

There are several numbers of issues that researchers should be aware of with respect to DA. These issues consist of:

Having the required skills to analyze the data;

Lack of clearly defined and objective outcome measurements;

- Improper subgroup analysis;
- Drawing an unbiased or fair inference;
- Way of presenting the entire data;
- Following acceptable norms for disciplines;
- Simultaneously selecting the methods for data collection and appropriate analysis;
- Providing honest as well as correct analysis;
- Determining statistical significance;
- Environmental/contextual issues;
- Data recording method;
- Partitioning 'text' at the time of analyzing qualitative data;
- Training of staff that is conducting analyses;
- Reliability and Validity; and
- Extent of analysis.

1.7.1. Having Necessary Skills to Analyze

A tacit assumption of many investigators is that they have acquired training that is enough to demonstrate a high standard of research practice. The unplanned or unintentional scientific misconduct is possibly the result of poor instruction as well as follow-up.

There are several studies that suggested that this may be the case more often than believed (Nowak, 1994; Silverman and Manson, 2003). For instance, Sica found that enough training that is provided to the physicians in medical schools in the appropriate design, implementation, and evaluation of clinical trials is "abysmally small" (Sica, cited in Nowak, 1994). In

Fact, a single course in biostatistics is the most that are generally provided (Christopher Williams, cited in Nowak, 1994).

Preferably, the investigators should have the knowledge more than just a basic understanding of the rationale for choosing one method of analysis over the other method. This can help the investigators to supervise staff, in a better way, who conduct the process of data analyses and make informed decisions.

1.7.2. Concurrently Selecting Data Collection Methods and Appropriate Analysis

Although the methods of DA may vary by scientific discipline, the optimal stage for determining the correct process of data analyses occurs early in the process of research and it should not be an afterthought.

According to Smeeton and Goda (2003), "Statistical advice should be taken at the point of the initial planning of an investigation so that, for instance, the method that is used for the process of sampling and design of questionnaire is correct."

1.7.3. Drawing Unbiased Inference

The main aim of the analysis is to differentiate between an event that is occurring as either reflecting a true effect against a false effect. Any bias that takes place in the collection of the data, or the choice of method of analysis, will increase the possibility of drawing a biased or unfair inference.

Bias can happen in the process when recruitment of study applicants falls below the minimum number that is needed to demonstrate statistical power or failure to maintain a sufficient follow-up period that is required to show an effect (Altman, 2001).

1.7.4. Inappropriate Subgroup Analysis

When an investigator fails to show statistically different levels between treatment groups, investigators may resort to breaking down the analysis to smaller and smaller subgroups. This is done by the investigator in order to find a difference between treatment groups.

Although, it has been observed that this practice may not essentially be unethical, these analyses should be proposed before starting the study even if the intent or the objective of the study is exploratory in nature. The investigator should make this explicit so that all the readers should understand that the research is more of a hunting expedition instead of being primarily theory driven. This is the case if the study is exploratory in nature.

Although it has been observed that a researcher may not have a theory-based hypothesis for analyzing the relationships between previously untested variables, a theory will have to be developed in order to explain an unanticipated or unexpected finding.Indeed, in case of exploratory science, there are no a priori hypotheses. Thus, in exploratory science, there are no hypothetical tests. Although it has been observed that the theories can often drive the procedures that are used in the investigation of qualitative studies. Many times, occurrences of patterns of behavior that are derived from analyzed data can result in the development of new theoretical frameworks instead of determining a priori (Savenye and Robinson, 2004).

It is possible that various statistical tests could produce a significant finding by chance alone instead of showing a true effect. Integrity is compromised in the case if the investigator just reports those tests that are having significant findings and ignore those large number of tests that fails to reach significance or the tests with no significant findings.

There are several computer-based statistical packages. It has been observed that the access to computer-based statistical packages can facilitate the application of increasingly complicated analytic procedures. On the other hand, the improper or inappropriate uses of these statistical packages can result in abuses as well.

1.7.5. Following Acceptable Norms for Disciplines

Every single field of study has established its accepted practices for the process of DA. Resnik (2000) stated that it is prudent or sensible for all the investigators to follow the accepted norms. Further, Resnik stated that these norms are '…based on two factors:

(1)   The nature of the variables that are used (for instance, qualitative, quantitative, or comparative),

(2)   Assumptions about the particular population from which the entire data is collected (for example, sample size, random distribution, independence, etc.). If an individual makes use of unconventional norms, it is very important to clearly state this is being done, and to show the way in which this new as well as a possibly unaccepted method of analysis is being used, as well as how it varies from other more traditional methods.

For instance, Schroder, Carey, and Vanable, (2003) had put together their identification of the new and powerful data analytic solutions that are developed to count the data in the area of HIV contraction risk. These analytic solutions were presented with a discussion of the limitations of commonly applied methods.

1.7.6. Determining Significance

The conventional practice is to develop a standard of acceptability for statistical significance, with certain disciplines. However, it may also be proper to discuss whether the attainment of statistical significance has a true practical meaning, that is 'clinical significance.'

'Clinical significance' was described by Jeans (1992) as "the potential or the ability for the findings of the research to make a real as well as a significant difference to clients or clinical practice, to health status or to any other issue that is identified as a relevant priority for the discipline."

Clinical significance was defined by many experts. Kendall and Grove (1988) define clinical significance in terms of what happens at what time "… after the treatment, troubled, and disordered clients are not different from a meaningful as well as representative non-disturbed reference group."

Thompson and Noferi (2002) suggested that the readers of counseling literature should expect all the writers of authors to report either practical or clinical significance indices, or both practical and clinical significance indices, within their research reports.There are times when some authors fail to point out that the magnitude of observed changes that may be too small to have any clinical or practical significance. At times, a supposed change may be described in some detail, in such cases but the investigator fails to disclose that the trend is not statistically significant." Shepard, (2003) questions why authors fail to point out the magnitude of changes that are too small to have clinical or practical significance.

1.7.7. Lack of Clearly Defined and Objective Outcome Measurements

There is no amount of statistical analysis, irrespective of the level of sophistication will correct the objective outcome measurements that are poorly defined. Whether this is done by mistake or unintentionally or by design, this practice enhances the possibility of clouding the interpretation of findings. Therefore, it is possibly misleading the readers.

1.7.8. Provide Honest and Accurate Analysis

The basis for this problem is the urgency of decreasing the possibility of statistical error. Some of the common challenges consist of the exclusion of outliers, filling in the data that is missing, altering or else changing data, data mining, and developing graphical representations of the data (Shamoo, Resnik, 2003).

1.7.9. Manner of Presenting Data

Sometimes, the investigators may improve the impression of a significant finding by determining how to present the derived data (as opposite to the data that is in its raw form), which part of the data is presented, why, how, and to whom the data is shown (Shamoo, Resnik, 2003).

Nowak (1994) also noted that even experts or professionals do not agree in differentiating between analyzing and massaging data. In addition to it, Shamoo (1989) suggested that investigators maintain a sufficient as well as a correct paper trail of how data was manipulated for review in the future.

1.7.10. Environmental/Contextual Issues

It has been observed that the integrity of DA can be compromised by the environment or context in which the entire data was gathered that is face-to-face interviews vs. focused group. The interaction that takes place within a dyadic relationship (interviewer-interviewee) is different as compared to that of the group dynamic that takes place within a focus group. This is because of the total number of participants, and how they react to the responses of each other participant.

Since the process of data collection could be affected by the environment/context, researchers or investigators should take this into account at the time of conducting the process of DA.

1.7.11. Data Recording Method

The method in which data was recorded could also have an impact on the analyses. For instance, research events could be documented by:

- Recording video or audio and transcribing it later;
- The preparation of ethnographic field notes from a participant/observer;
- Either closed-ended survey or open-ended survey;

- Either a researcher or self-administered survey; and
- Requesting all the participants to take notes, compile as well as submit them to investigators or researchers.

While every single methodology that has been employed has some logic or rationale and advantages, issues of objectivity and subjectivity may be raised at the time of data analyses.

1.7.12. Partitioning the Text

Staff researchers or 'raters' make use of inconsistent strategies in analyzing text material during content analysis. There are some 'raters' who may analyze the whole comments. On the other hand, there are some other raters who may choose to dissect or analyze the text material by separating phrases, words, clauses, sentences or groups of sentences. Every effort should be made in order to reduce the inconsistencies that are found between "raters." This is done so that data integrity is not compromised.

1.7.13. Training of Staff Conducting Analyses

The main challenge to data integrity could happen with the unmonitored supervision of inductive techniques. Content analysis needs raters in order to assign topics to text material (that is comments). The risk to the data integrity may occur when raters have received inconsistent training, or the raters may have received earlier training experience(s).

Previous experience may have an impact on the way the raters understand the material or even perceive the nature of the analyses that are to be conducted. Therefore, one rater can give topics or codes to material that is significantly different from the other rater. Strategies that are used in order to address this would consist of clearly stating a list of procedures used in analyses of the protocol manual, consistent training, and routine monitoring of raters.

1.8. CONCLUSION

DA is a term that is prevalent and important in modern times. DA describes the entire process of mining, cleaning, and analyzing the data. The main aim to perform DA is to give predictions for the development and establishment of the business. It is important to note that with the establishment of the internet as one of the major forms of business, DA has turned out to be one of the most effective of business establishment. There are several types of DA. These types are used as per the type of data given and the type of result that is desired.

While DA is performed for idealizing the business, these are several issues and considerations that come in the data given for analysis. Some of these major considerations are lack of clear identification of objects in data and the manner in which the data is presented. These are some new world problems that are pestering every data analyst around the world. It should be kept in mind that while performing DA, it is necessary to have a clear objective and the way the data is presented must be clear. Having clarity on these two issues will surely help in obtaining clearer results by DA.

REFERENCES

1. Bhasin, H., (2019). 8 Types of Analysis in Research—Types of Research Analysis. [online] Marketing91. Available at: https://www. marketing91.com/types-of-analysis-in-research/ (accessed on 10 March 2020).

2. Bhatia, M., (2018). Your Guide to Qualitative and Quantitative Data Analysis Methods—Atlan | Humans of Data. [online] Atlan | Humans of Data. Available at: https://humansofdata.atlan.com/2018/09/ qualitative-quantitative-data-analysis-methods/ (accessed on 10 March 2020).

3. Guru99, (2020). What is Data Analysis? Types, Process, Methods, Techniques. [online] Guru99.com. Available at: https://www.guru99. com/what-is-data-analysis.html (accessed on 10 March 2020).

4. Introduction to Data Analysis Handbook, (2006). [ebook] Migrant & Seasonal Head Start Technical Assistance Center: Academy for Educational Development. Available at: https://files.eric.ed.gov/ fulltext/ED536788.pdf (accessed on 10 March 2020).

5. Mohiuddin, A., & Al-sakin, K. P., (2019). Data Analytics, Concepts, Techniques, and Application. [ebook] CRC Press: Taylor & Francis Group. Available at: https://www.crcpress.com/Data-Analytics-Concepts-Techniques-and-Applications/Ahmed-Pathan/p/ book/9781138500815 (accessed on 10 March 2020).

6. Plato.Stanford.Edu. (2016). Analysis > Definitions and Descriptions of Analysis (Stanford Encyclopedia of Philosophy). [online] Available at: https://plato.stanford.edu/entries/analysis/s1.html#1 (accessed on 10 March 2020).

7. Sandra, D., (2019). Learn Here Different Ways of Data Analysis Methods & Techniques. [online] BI Blog | Data Visualization & Analytics Blog | datapine. Available at: https://www.datapine.com/ blog/data-analysis-methods-and-techniques/ (accessed on 10 March 2020).

Digital Marketing and Analytics

CONTENTS

Digital Marketing is the new age of marketing. This helps in reaching a large audience by investing fewer resources as compared to the earlier times. This chapter explains the various insights of digital marketing. The importance of digital marketing analytics along with various factors that affect digital marketing has been discussed in this chapter.

There are some important factors that must be considered by every organization or individual who is trying to digitalize their business. It has become very important to match the steps of the evolving world and get active on digital platforms.

This can only become possible if the tools of digital marketing are used properly. Also, this chapter explains the future of digital marketing analytics and how this will turn out to be in the future by matching the evolving technologies.

2.1. INTRODUCTION

It has been observed that today's era of the Internet has opened a gate of a vast range of opportunities for large and small enterprises. An individual cannot just share a private picture of one's birthday, but they can also get customers for their own business and reach them conveniently. This is made possible by the use of social networks. The speed and ease with which the digital media communicates information and give a boost to a business are just amazing.

Figure 2.1. In the modern world, Internet has turned out to be the part and parcel of life.

Source: Image by Needpix.

The term digital Marketing is used for the measurable, targeted, and interactive marketing of products or services by the use of digital technologies to reach the audiences, turn them into consumers, and retain them.

The traditional manner of marketing involved all the businesses, whether small or large, to promote their products or services on radio, print media and television commercials, business cards, bill boards, and in several other similar ways where Internet or social media websites were not employed for the purpose of promotion. Traditional marketing policies had restricted customer reachability and the scope of driving the purchasing behavior of the customer.

The following table lists a few points that distinguish digital marketing from traditional marketing

Traditional Marketing	Digital Marketing
• Communication is unidirectional. This means that a business communicates about its products or services with a group of other individuals. • The medium of communication is generally letters, phone calls, and Emails. • Campaigning takes more time for the process of designing, preparing, and launching. • It is carried out for a particular audience throughout from generating campaign ideas up to selling a product or a service. • It is a conventional way of marketing. It is best for reaching the local audience. • It is difficult to evaluate the efficiency of a campaign.	• Communication is bidirectional. The customer also can ask any kind of questions or make suggestions about the products and services of the business. • The medium of communication is mostly through chat, social media websites, and Email. • There is always a quick way to establish an online campaign and carry out changes together with its development. Campaigning is easier with digital tools. • It is best for reaching the audience from all over the world. • It is easier to evaluate the efficiency of a campaign through analytics.

2.1.1. Social Media Marketing

Social Media Marketing is defined as the way of generating website traffic or appealing to the viewers as well as customers through social networking websites. These websites include Pinterest, Facebook, LinkedIn, Twitter, and so on. In addition to it, social media marketing is a subgroup of digital marketing.

All the social networking websites are not necessarily employed for the purpose of digital marketing, but all these websites support the sharing of content. While there are some websites such as Facebook that emphasizes on personal sharing, Twitter emphasizes on tweeting short messages about the opinions or reactions of an individual. On the other hand, LinkedIn emphasizes professional networking and Pinterest motivates to market the ideas of an individual and online businesses.

2.2. DEFINING DIGITAL MARKETING

The application of the Internet and other kinds of digital media and technology to support 'modern marketing' has given rise to a bewildering range of labels and jargon. These are created by both academics as well as experts and professionals. It has been called by several names, which include, internet marketing, digital marketing, e-marketing, and web marketing. And also, all of these alternative terms have varied through time.

2.2.1. Digital Marketing Terminology

Digital Marketing is the term that is used most frequently in the present times. The frequent and prevalent usage of the term digital marketing is clearly visible from the definition. So, this is the term that is mostly in focus in the present times.It has been observed that there have been several discussions and debates conducted in order to decide the final and actual definition of the term 'digital marketing.' So, there have been several attempts, by several researchers, to finalize the exact definition of the term digital marketing.

2.2.2. Do Definitions of Digital Marketing Matter?

These definitions matter, since specially inside an organization or between any business and its clients there is a need for clarity in order to support all the goals and activities that support Digital Transformation. Also, it has been observed that many of the other definitions of digital marketing are misleading.

In the simplest form, digital marketing is defined in some of the book as simply:

"Attainment of marketing objectives through applying digital technologies and media."

In practice, digital marketing comprises managing different kinds of online company presence and presences like the website of the company, mobile applications, and social media company pages. All this is in conjunction with online communications techniques. This includes the likes of search engine marketing, online advertising, social media marketing, e-mail marketing and partnership arrangements with several other websites.

These kinds of techniques are useful in supporting the objectives of attaining new customers and offering the best services to the customers that are already existing. This will help in developing the customer relationship through E-CRM (electronic customer relationship management) and marketing automation.

Though, for successful digital marketing, there is still a need for integration of these techniques with traditional media, which includes TV, print, and direct mail as part of multichannel marketing communications.

If an individual introspects the other definitions of digital marketing from different sources like the definition of digital marketing from SAS: What is Digital Marketing and Why does it matter? or several other definitions of digital marketing from a number of other sources.

It is also evident that often the focus of this definition is on promoting products and services by the use of digital media rather than a more holistic definition which is covering the experience of the customer, relationship development and highlighting the importance of multichannel integration. So, the scope of the term should involve all the activities across the customer lifecycle:

- Digital and mobile experiences like web design and mobile apps;
- Digital media and different channels of communication;
- Digital strategies and integrating experiences and multichannel communications;
- Prospect and customer relationship management (CRM) through marketing automation; and
- Digital technology and platforms in order to manage all digital marketing activities.

The role of digital platforms in supporting integrated multichannel marketing is considered as an important component part of digital marketing. Yet, it has been observed that is it is often ignored. In several different ways, this emphasizes how important it is to break down silos between 'digital' and 'traditional' marketing departments.

The online channels can also be managed properly in order to support the entire process of buying from pre-sale to sale to post-sale and it is also useful in the further development of customer relationships.

2.3. UNDERSTANDING THE IMPORTANCE OF DIGITAL MARKETING

It has been observed that over the years, the business world is evolving continuously. In a similar manner, there have been several observations that have proved that marketing is not what it used to be in the previous years. In the present times, it has grown to be a lot more advanced and has become more target specific.

This means that marketing, particularly digital marketing, has become highly result oriented. And there is a presence of a great deal of data to work with, to understand the significance of digital marketing analytics.

For example, there was a time in the history technology when mail-order music services were in a great demand. In the present times, it has been substituted by technology focused innovations such as iTunes. It is useful for all musicians. They are promoting their music like never before. This was not possible in the past.An individual is required to understand that consumers no longer care about the generic approach irrespective of what kind of business he or she runs. They need a personalized experience to the core. Because of this reason, access to the correct kind of data is very important for all the marketers.

Everyone is living in an information age where data is considered as the new oil. Even if an individual is running a small business, there are high possibilities that one has come across the term "Big Data." Also, when it comes to business, it is important to understand the relevance and significance that it holds to an individual. It becomes important, for an individual involved in a business, to use technology and leverage data-driven marketing in order to get maximum return.

A digital marketer should always keep it in their mind that the right data at hand and knowing how to analyze it is very important. An individual can prove to be a game changer for their own business by having all the knowledge of digital marketing analytics and by using them in an effective manner.

2.3.1. Starting with the Basics of Digital Marketing Analytics

The ability to track and analyze the outcomes with the help of digital analytics is considered as one of the biggest advantages of conducting business in the digital era. Though, digital analytics is something that is different from simple web analytics metrics that one sees in the Google Analytics tool.

Web analytics is just a part of digital analytics. Web analytics is more on the metrics of the website. On the other hand, digital analytics allows an individual to have a more comprehensive view of the entire marketing strategy.

With the help of digital analytics, an individual comes to know what is working and what is not. Therefore, it provides a chance for an individual to improve their marketing plan. In addition to it, it also gives an individual the needed vision to see where they are actually going and if their attempts or efforts are paying off. This is a kind of luxury that one does not enjoy with traditional marketing. This is because there is no such option.

Whether it is running email marketing campaigns or just analyzing the visits on the website, there is a vast array of information that an individual can decode in order to understand their rate of success.

There are several numbers of important components that become more of a necessity than a mere need at the time of building and running a successful business in any niche. One of these components is having a website. This is because websites have become an important part of the modern-day marketing ecosystem.

An individual will not have access to critical analytics data if there is not a proper website in place. This would result in a lost business that cannot be recovered.

Those days are gone when customers just used to visit the websites of the business in order to get in touch with a company or gain more information about it. In the present times, all the individuals are using websites in order to converse with brands, make product inquiries, buy the required products while consuming targeted as well as appropriate content.

The websites of the business have become a digital marketing tool. It has become hard to ignore these business websites. This is because it helps not just to deliver the information but also the engagement of a high level. And in exchange for this, the business receives is some important data and insights. These crucial data and insights can be used in order to improve marketing efforts and to further grow the business.

2.3.2. Blog Comments

Having a company blog can be a very useful or a game changer for all those brands that want to make the most of content marketing. In the context of delivering value to the customers, the content of the blog is known to help all the businesses to build a brand and collect all those targeted leads that convert.

Though, there is one thing that an individual may ignore. It is the fact that the comments on the blog are a great way which helps in understanding the audience and what they are actually looking for. There is no need to even have an analytics plugin installed to see how the audience is reacting to all the content on the blog.

The individual has the chance to gauge the interest of the audience when he or she post content on the blog.

- Is the audience reacting positively to the content? Or they are responding negatively?
- Are there times when the person experiences zero engagement from the readers?
- What kind of readers is posting the most comments on the content?
- Is the audience asking insightful questions that one can provide an answer to?

The proper understanding of digital marketing analytics is not only about looking at the core numbers. But is also about evaluating and getting a clearer insight into the requirements of the customers/prospects by looking at the behavior of the customer. For instance, if there is the presence of few ideas that often get mentioned in the contents of the blog, then one may want to analyze the ideas to see if there is something important for them.

Creativity is an important element when it comes to the collection of data and knowing the customers. So, it is the responsibility of an individual to go beyond comments and evaluate how many shares the blog post is getting. An individual should think out of the box and it will become easier for them to grasp the right digital marketing analytics data and use it in an efficient manner.

2.3.3. Google Analytics

There exist a few things that lay the foundation when it comes to digital marketing analytics. Google Analytics is one of them. The type of insight that an individual derives from Google Analytics may not be too costly, but it does create the path for understanding the data that is available and making sense of that data.

There are many different kinds of tools that are used for the collection of data (both advanced as well as basic). Few of these tools are as robust as Google Analytics. Being a robust and reliable tool makes Google Analytics one of the commonly used tools in modern time.

Google Analytics provides a set of tools that are not like any other set. This fact is the good thing about Google Analytics. Whether an individual is a novice or a professional, one can easily look into the data that is found in it not only to understand the target demographics but also to figure out the best possible way in order to reach out to them.

Now, if an individual is new to a platform such as Google Analytics, this platform may not seem as easy or simple as it sounds to everyone. Initially, the data that an individual gets to access can be particularly overwhelming and it may also look little confusing to them.

However, if the website of an individual is not that old then one can do some surface-level evaluation. This evaluation is done in order to understand the needs of the customer with the help of some basic metrics that throw light on the performance of the website.Or if in case an individual is not good at analyzing the data that is available, then he or she can always hire a digital marketing professional. The professional can help in understanding what the data is actually telling and what should be done about it.

2.4. DASHBOARD METRICS

The proper understanding of the performance of the website begins by looking at the important metrics that are shown on the application's dashboard. It provides a bird's view to an individual about how the website is growing and a basic understanding of what one can do in order to enhance the numbers.

Out of all the statistics that one can see on that dashboard, there are three important statistics that an individual should pay attention to. These are as follows:

Figure 2.2. Various aspects of dash metrics.

2.4.1. Pages Per Visit

This is the metric that allows an individual to know the number of average visitors that have viewed the pages of the website before they left. The main purpose is to increase this number of average visitors. This can be done by improving the content that is posted on the website and on-site value. One should see a high number by keeping all the things appropriate and of high quality on the site.

2.4.2. Average Visit Duration

When some person visits any type of website, they basically see for the things they are interested in. Whether the content is more information about the product or content is associated with the business. The task of increasing this number should be on the daily to-do list of an individual.

This is because having a low number shows that one did not provide enough value to the people who are visiting the site and viewing the content. There is a direct relation between the business output and average visit duration. If an individual is able to attract visitors to spend more time on their website, there are more chances that the business of that individual will surely rise.

2.4.3. Bounce Rate

It has been observed that people may land on a particular website from different points of entry or there are several sources to visit and view the website. For instance, these visitors may find the homepage or the blog post.

But once these visitors have found the site, it is the responsibility of an individual to ensure that they explore the entire site and take some action. The bounce rate of the website shows the percentage of people that left the website after visiting a single page.

Having a high bounce rate is not necessarily bad but one should always try to lower the bounce rate. This is done by improving the internal navigation of the website and by providing better content on the site.

By looking at all the three metrics that are discussed above, one can learn more about how the people interact with the website once they have found it. And, what else can be done in order to improve the website to make it more appealing to the visitors.

2.4.4. Website Traffic

Website Traffic is defined as the amount of traffic or the individual who is coming to the website. It is essential to check the traffic of the website on a daily/weekly/monthly basis. This is totally dependent on digital marketing goals. Checking the traffic of the website will help an individual fine tune the digital marketing strategy on a regular basis.

2.4.5. Sources of Traffic

It is equally important for an individual to understand the sources of traffic on the website. This helps an individual to find out which platform should be made use of more than others, to advertise the business.

2.4.6. New vs Returning Visitors

New vs returning visitors is the other parameter to check if there are some new visitors to the website and also to see the number of visitors those who are visiting the website more than once. It is very important to have increasing numbers for the section of a new visitors. This would mean that a greater number of individuals are being made aware of the business.

At the same point of time, it is equally important to make sure the number of returning visitors are increasing and if in case this number is not increasing, at least it should not decrease. The increase in the number of returning visitors to the website would mean that the website has enough value to attract all the visitors to make a second visit to the website, and then a third, a fourth, and so on.

2.4.7. Average Time on Page

The average time that is spent on a page of the website will help an individual to understand how valuable the page is for the target audience.

2.5. THE 5DS OF DIGITAL MARKETING

In order to understand the significance of digital marketing to the future of marketing in any business, it is very important and helpful to think about what audience interactions one is required to understand and manage.

In present times, digital marketing is about many more kinds of audience interaction than just a website or email. It also includes managing and harnessing the '5Ds of Digital' that are defined in some of the books.

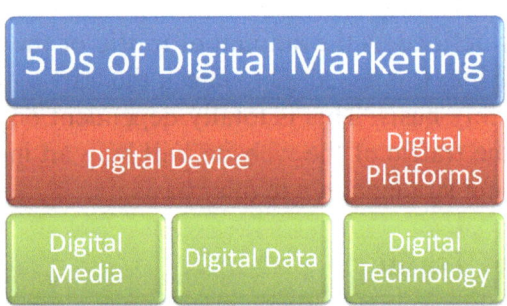

Figure 2.3. It is important to work on the 5Ds of digital marketing to gain more benefit.

The 5Ds of digital marketing describes the opportunities that are available for all consumers to interact with brands and for any kind of businesses to reach and learn from their audiences in several different ways:

2.5.1. Digital Devices

Audiences get the experience of a particular brand as they interact with business websites and any other mobile apps. The customers interact through a combination of connected devices which comprises tablets, smartphones, TVs, desktop computers, and other gaming devices.

2.5.2. Digital Platforms

Most of the interactions by the customers on these devices are done through a browser or apps from the main platforms or services. This includes Google (and YouTube), Facebook (and Instagram), Twitter, and LinkedIn.

2.5.3. Digital Media

These are the different kinds of paid, owned, and earned communications channels that are useful for reaching and engaging audiences which comprise email, advertising, and messaging, search engines and several other social networks.

2.5.4. Digital Data

The insight businesses gather the information about the profiles of their audience and their interactions with the businesses. In the present times, this is required to be protected by law in many of the countries.

2.5.5. Digital Technology

The marketing technology, also known as Mar-Tech, is very important and makes it clear that businesses use these technologies to create interactive experiences from websites and mobile apps to in-store kiosks and email campaigns.

2.6. PRINCIPLES OF DIGITAL MARKETING

2.6.1. Content Capital

With the presence of seemingly endless quantities of content on the sites, it probably comes as no surprise that a troubling amount of this content that is available is total garbage. It has been observed that effective digital marketing will set itself apart from this garbage. This is done by resting its laurels on engaging, solid, as well as sharable content.

The Internet is free for everybody and it is available to everyone, so anyone can (and it can seem like everybody does) collect their own content. Also, it has been observed that not all of this content is quality. There have been several instances where it has been observed that content is not of proper quality and does not match the expectations of the customer.

So, in this scenario, it is important that customers must the content as reliable and credible. It is very important for any individual to present reliable content that has some credibility and matches the expectation of the customer.

2.6.2. Simplicity

Simplicity is one of the major aspects of digital marketing, the simpler the message, the more efficient it will be. Most of the people thoroughly connect with the digital marketing message that is simple and conveys the required information. It has been observed that amazing digital marketing initiatives miss the mark simply. This is just because they got a little overzealous and overcomplicated.

Therefore, it is advised to keep the message succinct. All the individuals have enough going on in their individual lives. When something is provided in a simple manner, customers will revel in its simplicity.

2.6.3. The Customer Connection

An individual must know about the customer. This digital marketing principle may seem very basic, but often it is lost in the struggle to make the next 'big' idea. The best angle is one that attracts the attention of the customer by just speaking in a direct, clear, and simple manner.

An individual should make sure that the digital marketing campaign targets the specific audience. While curating the content, the business owner must not go too broad. Wandering off from the main topic or content could inflict losses on your targeted market.

2.6.4. Pimp Your Vitals

It has been observed that the majority of visitors visit the website in order to get the most basic of basic information, such as street address, location, email address and phone number. It is the duty of an individual to make sure that all these important details are super easy for the visitors to find.

2.6.5. Buddy Up

One should make beneficial alliances with other online businesses and initiatives. This can help in promoting and strengthening the online presence, the offline presence, as well as the digital marketing efforts. This does not mean that an individual should slide into bed with their direct competition, rather one should find other brands that are available, that compliment your brand. (Think McDonalds and Disney or Loblaw's and Esso.).

2.6.6. Video

Last year video got set to steal the stage, and this year it's done just that. An individual should make some use of quality video marketing content in order to improve the digital marketing message. There were several studies done that show when someone sees and hears a message, the ability of that person to recall that particular message increases exponentially.

2.6.7. Media Medley

Social media is very important. But one should not forget about old school media. It is equally important to remember the 'old school' media such as broadcast, press, and (gasp!) even print. Embracing all kinds of media will help in creating a cohesive, solid as well as well-branded foundation for all the customers.

2.6.8. Face Value

Everyone is aware of the fact that beauty is only skin deep, but the 'face' of the company (and consequently, the digital marketing) is required to look at its best.

2.6.9. Consistency Is Key

The relationship with the customers is not just a one-way street; like any other relationship. The relationship with the customers goes both ways. And, also if an individual wants to ensure long-term success then the individual must be consistent in their behavior and customer interaction.

2.6.10. Adapt or Die

The digital world is evolving continually. As a result of this, digital marketing is also evolving. The individuals are required to commit themselves to constantly aiming to enhance and expand. An individual should keep resources such as time and money away for digital marketing R&D.

One can attain the best results if they set aside almost 10% (and no less than 5%) of their time dedicated to research. If any individual does not want to waste time in hopping from website to website, then they must put the time and effort in identifying the trends or forming their own search. Customers usually identify if they are reading the copied content. And this downgrades the image of the organization that is publishing copied content.

2.6.11. Act on Feelings and Instinct

It is well known that more than people buy the products, they buy feelings. The best example to explain this phenomenon is the powerful marketing used in alcohol commercials. The advertisements for alcohol are perhaps the best example of such precise and effective marketing. These advertisements are majorly aimed at getting people to purchase a product by feeling about the product in a certain way. They are forced to believe that only a particular product is capable of helping them achieve their desired lifestyle.

2.6.12. Know the Competition

An important area of research and development must involve recognizing, observing, and analyzing the competitors. A particular organization or website can grow in the modern world only if it has proper knowledge of the activities of its competition. This helps the organization in getting the bigger picture of what and where that organization or website must be employing their digital marketing efforts.

2.7. DIGITAL MARKETING ANALYTICS TOOLS

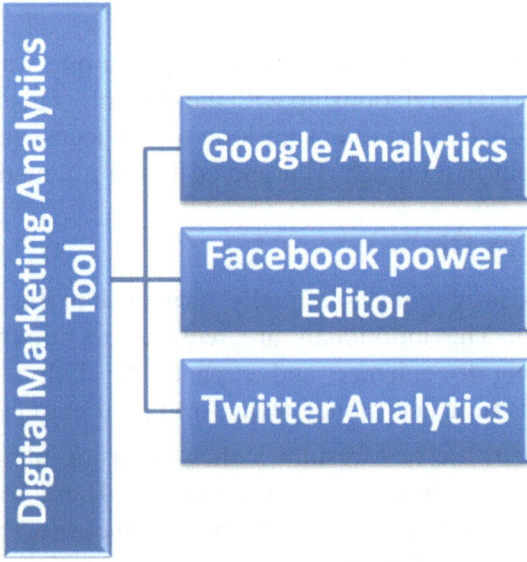

Figure 2.4. Digital marketing analytics tools.

2.7.1. Google Analytics

Google Analytics helps an individual to monitor the analytics that is associated with the website. In addition to it, it also helps in monitoring the other parameters that are related to the website. All that individual has to do is integrate the code that is provided by Google Analytics the website. And once the website is implemented in a proper manner, the individual will be able to track their traffic on the website easily and fine tune the website promotion strategy.

2.7.2. Facebook Power Editor

This is a tool that not just helps an individual to run paid ads on Facebook, but it allows them to monitor the campaigns. This will help an individual to fine tune the campaign based on the performance.

2.7.3. Twitter Analytics

Twitter Analytics helps the individual in understanding the performance of the tweets. Also, it will help them to figure out the kinds of tweets that work best for their industry and also for the entire target audience.

2.8. WHY DIGITAL MARKETING ANALYTICS MATTER TO YOUR BUSINESS

In the year 1955, Columbia House sold 700,000 records with the help of direct mail marketing and a telephone ordering service. In the present times, 93% of marketers agree to the fact that social media is essential for the business.

It has been observed that the business world is changing very quickly, and along with it there is a rapid change in marketing. Columbia House and some other mail-order music services provided way to giants such as Sam Goody and F.Y.E. Few years later, this provided way to iTunes.

If this kind of trend demonstrates one thing, it is that the consumers in the modern times need a personalized, convenient experience that meets all their requirements specifically, not just the generalized expectations of a homogenized mob.

Data is changing the game in the information age. The term Big Data is heard by the marketers of both large as well as small businesses. But what does it actually mean to the business that they are running? Data is becoming a need for all marketers, mainly for the inbound marketers, and the insights that are provided are essential, profound, and these insights are readily available to those individuals who know where to have to look.

2.8.1. The Need for Data

Data and its place in business is a growing discussion. For years, it has been observed that business has been a speculative venture, to an extent. There are no direct tracking mechanisms for ad campaigns. The collection of customer demographics and information about their lifestyle needed low-response-rate telephone efforts and manual data collection.

Even when the data was available, the mechanisms for examining, comparing, and utilizing that information were not adequate. In fact, at that time, there was no technology that could make the process of examining, comparing, and utilizing worthwhile.

It has been observed that the larger companies had the resources that were useful in scraping the data from their customer base. On the other hand, small businesses were forced to function in a kind of ecosystem that was opaque by today's standards.

Flash forward, past the birth of the personal computer and the construction of vast and powerful computer networks: the game has changed.

The analysis and understanding of the behavior of the customer, marketing efficiency, and business strategy are available to large and small enterprises by massive data collection efforts and equally massive computational capabilities.

This alone is sufficient to merit the requirement for data. Whenever there is an opportunity for businesses to include better products and services offered by technology, they must grab it. Though, this motivation alone pales in comparison to the increasing expectations of the customer in an increasingly enabled environment.

It has been observed that 25% of consumers who make a complaint to a company on social media expect a reply within 1 hour. Emails that were triggered by specific customer behaviors produce 71% higher open rate. 102% higher click rates were produced than non-triggered, boiler-plate email messages.These statistics paint a simple picture: customers in modern time needs more responsive and personalized service through those channels that are more comfortable to them. Also, it has been observed that there is a long hold times on customer service lines. And these not acceptable anymore.

Mass mailings as well as copy/paste marketing are no longer converting when the division of audience provides the kind of small-business service that customers have grown to expect.

The analysis and collection of data, particularly in volumes enabled by the Internet, offers a greater insight into the lives, preferences, and needs of the customer. In addition to it, it also provides greater insight into the effectiveness of business practices as well as strategy. This also offers actionable, concrete information on which to base essential decisions in a myriad of business contexts.

2.9. HOW TO USE ANALYTICS IN DIGITAL MARKETING

Digital marketing is all about getting the individuals to view and interact with digital content. In traditional forms of marketing which include posters, mail, billboards, and TV commercials, it can be exceedingly hard to evaluate the efficacy of the message.

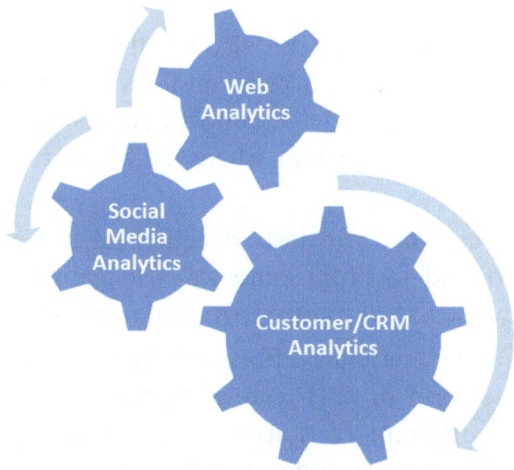

Figure 2.5. Three main components of analytics in digital marketing.

However, every single view and interaction is available as raw data when marketing goes digital. If an individual exactly knows how the digital marketing campaign is doing based on any number of metrics, it will help in providing an actionable answers to the individual as to what to enhance and what to preserve.

2.9.1. Web Analytics

It has been observed that the end goal of most of the most digital marketing efforts is to have a visit of the audience on the site. It eventually converts by purchasing anything, filling out a form that is available, or picking up the mobile phone.

Simple web analytics tools help an individual to see what happens exactly between visiting the site and converting, or, perhaps more vitally, not converting.In addition to it, this tool allows an individual to count the visits on the site, page-views, how visitors move from one page to the other page, how long the visitor stays on the particular page, and how much of the page the visitors view. On the surface, these observations allow an individual to confirm the efficacy of the design, layout, and content of the site.

An individual is required to consider redesigning the navigation and layout of the site if the visitors on the site are not following the planned flow through the internal links that are available on the website. In a similar way, one can see if visitors are wrongly identifying the primary content of the page, and one can learn from the successful content on how to enhance what is missing.

Going deeper, an individual can also see the times of day and days of the week when one gets the most visits on the site. They can also compare those to the times when they get the most conversions. This will provide some insight to the individual into where the highest converting traffic is found on the site.

An individual can also run tests designed in order to focus on non-converting visitors and identify some ways to gain the interest of these visitors (maybe by the use of longer-term lead nurturing strategy).

This type of web analytics effort can be used in various real-time experiments in the form of A/B testing. In this type of testing, different visitors are shown different versions of a particular site or feature of that particular site. This is done in order to collect statistics on how the visitors behave when they see a particular kind of feature.

These tests can be run on all the parts of the site by an individual. An individual should choose that versions of the site that are most efficient for the conversion rate. Large web entities such as the Huffington Post usually perform A/B testing even on their headlines, which allows the individual to know within some minutes about the version which has a higher conversion rate.

2.9.2. Social Media Analytics

Social media provided an opportunity for the viewers to do some marketing for the site and provides some great data on engagement. This data is provided by the viewers in the form of Facebook likes, tweets, social shares, and more.

If an individual has a website that includes blog, then social media may be the major avenue for promoting all the new posts, or the metric for how effective an article was. There should be proper analysis and comparison that must be made between the quantity of social media interaction for each post to the detailed metrics about the actual article (such as the post length, header/subheader layout, use of images/videos, writing style, etc.). This analysis and comparison will surely help in determining what kind of features are best for interacting customers, and thus proceed further by outing the same kind of efforts.

Social media may also be considered as a way for an individual to gain more traffic and advertising the brand. In this case, an individual wants to monitor retweets as well as shares, and they also need to look into the emerging sentiment analysis market.

On a large scale, this will permit an individual to monitor whether mentions of the brand are generally positive or these mentions are negative. Also, it will become easier to notice the other things that the users usually refer to a particular website in the context of (which can be the start of a targeted marketing campaign).

2.9.3. Customer/CRM Analytics

The main aim of marketing the content is to draw in the customers. Every information that is received by the customer is a data point. This data can be analyzed in order to learn about which people are getting influenced by the marketing strategies and which are not being impacted by these marketing strategies. Also, this provides an insight of how the improvements can be brought to gain more benefits.

Customer relationship management (CRM) software is used to store the data. This is ideal for standardizing the data of the customer in just one location. Such kind of programs helps in making analytics far simpler than trying to combine multiple records.

It has been observed that many BI software come with analytics features that are built in. On the other hand, for those CRMs which are without this feature, it is still possible to use Excel (or a full business intelligence solution) to collect data from the CRM, web analytics, and social media interactions. This data is collected in order to find patterns and user correlations.

This is known as data-driven customer profiling. The main purpose of data-driven customer profiling is to show the number of the customers that are attracted to the site. Also, it shows the number of users on which are not finding the content of the website relevant.Once an individual gets to know about this number, one can either pivot the goals to cater to the already-receptive demographics, or he or she can re-evaluate the content and marketing strategies to find out where an individual got off track.The insight that is collected from these sources of data (that includes websites, social media, and CRM) can be used in order to enhance the digital marketing strategy, either as a whole, on every post on the blog, or even with every share through social media.

Writing as well as design skills alone can only take an individual so far. The users usually click on some things more often than others, and the creator or editor of the website is not able to know how all the users actually behave until one starts analyzing and collecting the appropriate data.

2.10. IMPORTANCE OF DIGITAL MARKETING ANA-LYTICS

Over the last few years, it has been observed that the significance of digital marketing has measurably expanded in most of the industry. The e-Commerce enterprise can significantly extend its reach, function in a more effective manner, and move closer to attaining its long-term financial objectives. This is made possible by the appropriate digital marketing strategy.

While the benefits of digital marketing are evident as soon as the efforts are put in, knowing if the team working on digital marketing has adopted a specific approach poses a much more difficult challenge.

It is considered much better for an individual to look for quantitative evidence that supports what they have been doing rather than looking for a kind of approach that "seems" to be paying off.

It has been observed that the world of digital marketing analytics has been able to keep up with the marketing environment that is changing quickly. There are found to be different angles that the e-Commerce business can look at what one has accomplished; the one angle that makes the most sense for the individual will depend on the particular nature of the business.

2.11. THE PRESENT AND FUTURE OF MARKETING ANALYTICS

Over the last decade, marketing analytics has evolved throughout. It has been used in various fields and for different objectives. It has been seen that throughout the years the capabilities of marketers have changed drastically.

This increasing complexity in marketing has also altered the expectations of the consumer and how they decide whether or not to involve a particular brand.

All these changes indicate a need to adapt through improved analytics that goes beyond all the models that were used in the past. In order to get these analytics and involve all the consumers across today's complex marketing mix in an efficient manner, marketers are required to understand the modern marketing landscape. In addition to it, it is also required by the markets to understand the expectations of the consumer that influence it.

2.11.1. Modern Marketing and the Omni Channel Landscape

The digital revolution brought with it a completely new media mix have for marketers. This helps all the marketers to leverage in order to reach the consumers. In the present times, everyone has access to a wide range of mobiles, desktop, IoT, and several other devices that offer opportunities for brands to connect their products as well as services with people all over the world.

As a result of this, the marketing landscape has grown complicated, providing more options to the customers for where and when they decide to get engaged with brands. For all the marketers, this places a major focus on the necessity to have the right message, on the right channel and at the right time.

It is not only the digital landscape that has developed. In the present times, as more and more channels have been introduced, print as well as broadcast media has also grown complicated. For instance, during the period of 50's and 60's, there were just three main television stations, ABC, CBS, and NBC.

In previous years, there were 1,761 stations. This means that in the present times, there are more touch points across a more diverse media landscape with several other brands that are competing for the attention of consumers through each channel.

The presence of an excess of engagement options for consumers, together with the approximate 10,000 advertisements that are placed in front of them every single day, has given rise to a new, consumer-empowered marketing landscape.

In order to increase marketing efficacy in this environment, marketers are required to understand the expectations of the target customers.

2.11.2. Changing Consumer Expectations

Everyone is aware of the fact that consumers in the present times have a wider variety of media channels from which they can be reached by some brands. In addition to it, they also have the capability to shop online. The entire credit goes to eCommerce. These consumers have come to expect more from those brands that are trying to reach them.

In the present times, consumers also want to be associated with the products and services that they require, the minute they need them. For all the marketers who are looking to meet the expectations of the customers, they are required to understand the behavior of the consumer down to an individual, person-level.

From this point, they can start to understand the unique interests and pain points that the consumer has, the channels that they wish to use across the marketing mix, and the creative messaging that sets a brand apart from all the competition.

In addition to it, consumers also expect brands to be able to personalize the experience of the customer across all channels—online as well as offline. This expectation of the consumers for appropriate, timely, and personalized omnichannel engagements mean that all the marketers are required to again shift the kind of ways through which they measure their efforts. In the present times, marketers require person-level data that links together the unique activities as well as the experiences of an individual.

2.11.3. Traditional Attribution Models Are Obsolete

As digital emerged as the predominant channel leveraged by marketers, Digital attribution models play an important role for the marketers. The use of digital attribution models such as multi-touch attribution provided ways to marketers to quickly understand the touchpoints as well as channels that played the most important roles in driving revenue.

Subsequently, this permitted all marketers to improve their campaigns. This is done by giving priority to the channels that drove the highest engagement.

Though, this method is faulty. In the present times, revenue is not only driven by digital marketing efforts alone, but instead it required a combination of offline media, online media, and external factors. All this have an impact on the consumers and encourage them to make a purchasing decision.

In the present times, markets are required to do more in order to understand marketing efforts across all these channels and improve them to meet the expectations of the consumers. In Particular, there is a need to implement a unified marketing measurement strategy.

2.12. THE FUTURE OF ANALYTICS AND HOW IT'S CHANGING DIGITAL MARKETING

It has been observed that data science and machine learning (ML) are rapidly changing every industry. Starting from the fashion industry to finance, these technological capabilities are inspiring all the users by providing them the chance to attain more with the time.

Data scientists were used to spent about 80% of their time doing prep work before the technology of today. But now, calculations that used to take long times such as weeks, and sometimes it takes even months, can be calculated instantly.

The ability for the owners of the business to see, understand, and use their analytics for the development of their brand has never been simpler. And this is the only reason why the future of the analytics industry is so bright.

Businesses can make more informed decisions regarding their digital strategy with the help of data analytics tools. This will help in providing them an edge over their competition.

Analytics has always played a significant role in digital marketing and it will always play. This is because it helps businesses to make a data-driven marketing strategy. Marketers are required to know where their money is actually going and what the return to their business will be.

Whether it is like, shares on the site or conversions, the future of the analytics industry together with artificial intelligence (AI) and data science is changing the ways how marketers perform their jobs.

2.12.1. Diagnostic Analytics

When there is the presence of enough historical data to completely understand why something has happened, diagnostic analytics comes into play. For instance, if an individual ran an ad campaign. He did some country setting in the background, and sales did better in suburban areas as compared to that of urban areas. All this information will be revealed to that individual by diagnostic analytics. This is essential because one can now begin a retargeting campaign or segment the ads in order to cater to various audiences.

2.12.2. Descriptive Analytics

Descriptive analytics is defined as the data that is gathered during the preliminary stages of data collection and processing. It is a raw data that shows an individual where the users are going on the internet and what they are actually doing there and somewhere else.

With ample time and a huge amount of data, this information turns historical data for the business. Later, this can be used as a reference which helps an individual in making more informed decisions about the future of data analytics for their business.

2.12.3. Predictive Analytics

The term predictive analytics is just what it sounds like: predictions. This data represents what is most likely to happen which is based on descriptive and diagnostic analytics. In addition to it, it also helps an individual to estimate how sales will do in various regions, demographics, and segmented viewers.

An individual is able to see how campaigns could play out before even launching them with the help of predictive analytics. This is useful because the team can alter and change things until every single thing is perfect.

2.12.4. Prescriptive Analytics

Prescriptive analytics is what an individual is going to prescribe or recommend to the problem. An individual is required to fix all the weak spots that are based on all the data that has been gathered about the sales cycle. At this point in time, ML and augmented analytics comes into play. These technologies can help an individual to map out their next moves.

2.12.5. Artificial Intelligence and Machine Learning

Augmented analytics is one of the most efficient technology-powered solutions. It can help in transforming business ideas into successful business capabilities. As the name explains, augmented analytics is the capability to evaluate, analyze, review, and interpret data. This is done with the assistance of AI and ML. Augmented analytics becomes very helpful as it provides the business owners to take actionable steps that they can use to enhance their business.

These tools not just help an individual in saving time and money, but it also provides more time for the individual in order to attend the other matters and help in growing the revenue. There are benefits to this for businesses of all shapes and sizes.

2.13. CONCLUSION

Digital marketing analytics has turned out to be one of the most important aspects of technology-driven time. Using analytics in digital marketing not only explains the customer behavior to the business owners, but also explains their shortcomings.

With properly using digital marketing analytics, any individual or organization can benefit their business. Digital marketing helps in explaining what works best for a particular business and how the strategies can be formed for future operations.

Also, digital marketing can prove to be very helpful for a particular business as it helps them in gaining knowledge about their competitors. There is a lot that can be done if the digital marketing analytics are used properly. Furthermore, as the human being steps into the future that is controlled by technology, it becomes very important to have thorough knowledge about using digital analytics properly.

REFERENCES

1. Chaffey, D., (2019). What is Digital Marketing? A Visual Summary—Smart Insights. [online] Smart Insights. Available at: https://www.smartinsights.com/digital-marketing-strategy/what-is-digital-marketing/ (accessed on 10 March 2020).

2. Digital Marketing Blog. (2019). Understanding the Importance of Digital Marketing Analytics. [online] Available at: https://www.lyfemarketing.com/blog/digital-marketing-analytics/ (accessed on 10 March 2020).

3. Firth, C., (2014). 13 Principles of Digital Marketing. [online] iSTORM New Media: Digital Marketing Blog. Available at: http://blog.istorm.ca/marketing/content-marketing/13-principles-digital-marketing/ (accessed on 10 March 2020).

4. Fisher, V., (2019). The Importance of Digital Marketing Analytics | CustomerThink. [online] Customerthink.com. Available at: http://customerthink.com/the-importance-of-digital-marketing-analytics/ (accessed on 10 March 2020).

5. Hebert, C., (2014). How to Use Analytics in Digital Marketing. [online] Brightpod.com. Available at: https://www.brightpod.com/boost/how-to-use-analytics-in-digital-marketing (accessed on 10 March 2020).

6. Marketingevolution.com. (2019). The Present and Future of Marketing Analytics: Person-Level Data Through Unified Measurement. [online] Available at: https://www.marketingevolution.com/knowledge-center/the-future-of-marketing-analytics-person-level-data-through-unified-measurement (accessed on 10 March 2020).

7. NGDATA. (n.d.). NGDATA | Digital Marketing Analytics. [online] Available at: https://www.ngdata.com/dictionary/digital-marketing-analytics/ (accessed on 10 March 2020).

8. Palmere, T., (2019). The Future of Analytics and How It's Changing Digital Marketing | State of Digital. [online] State of Digital. Available at: https://www.stateofdigital.com/future-of-analytics-changing-digital-marketing/ (accessed on 10 March 2020).

9. Raghunath, M., (2017). An Insider's Guide to Digital Marketing Analytics. [online] Digital Vidya. Available at: https://www.digitalvidya.com/blog/digital-marketing-analytics/ (accessed on 10 March 2020).

10. Single Grain, (n.d.). Why Digital Marketing Analytics Matter To Your

Business. [online] Available at: https://www.singlegrain.com/blog-posts/analytics/why-digital-marketing-analytics-matter/ (accessed on 10 March 2020).

11. Srivastava, T., (2018). A Comprehensive Guide to Digital Marketing and Analytics. [online] Analytics Vidhya. Available at: https://www.analyticsvidhya.com/blog/2018/12/guide-digital-marketing-analytics/ (accessed on 10 March 2020).

12. Tutorialspoint.com. (n.d.). Digital Marketing – Introduction – Tutorialspoint. [online] Available at: https://www.tutorialspoint.com/pinterest_marketing/digital_marketing_introduction.htm (accessed on 10 March 2020).

Introduction to Artificial Intelligence

CONTENTS

The chapter of Introduction to Artificial Intelligence mentions the basic significance of Artificial Intelligence as well as its impact on every-day life. This chapter also explains the key drivers of Artificial Intelligence, what is super-intelligence, general artificial intelligence, narrow artificial intelligence, and machine learning.

This chapter comprises the permeation and application of artificial intelligence and how artificial intelligence is transforming and revolutionizing the day to day life. This chapter provides highlights on the various application of the artificial intelligence such as artificial intelligence in marketing, finance, health care, automotive industry, and the impact of the internet of things with the artificial intelligence.

This chapter also explains how artificial intelligence roots to unemployment, what are the dangers of artificial intelligence. This chapter also mentions the role of planning in artificial intelligence, what are the advantages and disadvantages of Artificial intelligence.

3.1. INTRODUCTION

Artificial intelligence is a methodology or tactic to make a robot, computer, or a product to think or process how smart human thinks. Artificial intelligence (AI) is a study about how the human brain thinks, decides, learn, and operates, when it comes to finding the solution to any problem or issue.

And finally, this study, as an outcome, gives intelligent software systems. The prime objective of AI is to develop or enhance the functions of the computer which are related to human knowledge, as an instance, learning, reasoning, and problem solving.

The intelligence is intangible. It is composed of factors that are mentioned below:

- Reasoning;
- Learning;
- Problem Solving;
- Perception; and
- Linguistic Intelligence.

The main goals of AI research are the representation of knowledge, learning, reasoning, natural language processing, planning, realization, and the capability to move and manipulate the objects. There are long-term objectives in the general intelligence sector.

These approaches are consisting of computational intelligence, statistical methods, and traditional coding of AI.

Through the course of the AI research are related to the search and the mathematical optimization, artificial neural networks (ANNs) and the tactics which are based on probability, statistics, and economics, generally there are several numbers of tools that are being used. Computer science attracts the AI in the field of mathematics, linguistics, science, psychology, philosophy, and so on.

Figure 3.1. The introduction of artificial intelligence.

Source: Image by Pixabay.

In much simpler terms, the AI is the simulation of Human Intelligence with the help of the machines.

AI has had an extreme rate of development in the past several years. Wherever an individual belongs to, finance, development, management, design, marketing, and sales or support, that individual has a fear of the AI posing a threat. Let's dive into the AI to understand it in a much better way.

3.1.1. AI Is a Broad Concept

There are several numbers of aspects which make humans different from the machines, that is, thought, taste, vision, odor, speech, feel, and hear. In the present interval of time, engineers that deals with the Data Science and Machine Learning (ML), are working hard to recreate the sense of thought, speech, visions in machines and speech.

If the machines have the thinking capabilities like humans, then the machines can understand and speak in the same language as humans do, and also, can see what human see and process it in the same way, then it means that machines are trying to match the intelligence which is specific to the humans.

In this way, the machines can develop and improve the work of humans and make it much easy with the help of the automation, prediction, and elevated efficacy of the work with respect to the speed as well as the precision.

There are some examples that are given below, that explains it to some extent:

- Thought—Facebook uses artificial neural networks (ANNs). It has the algorithms that copy the structure of the human brain for facial recognition. It recommends a tag if someone uploads a photo that has the face of that person in it in line with Privacy Settings;
- Speech and hearing—Siri, Cortana, Google Assistant, Alexa, Bixby; and
- Vision—Industry surveillance, capturing traffic violations.

3.1.2. The Past

The oldest (and also it is still existing) the application of AI is the Autopilot in Flights that dates to 1914.

In the early phase of the aviation, aircraft necessitates a pilot to continuously pay his or her consideration to fly with full safety. Flights of several numbers of hours led to serious exhaustion with respect to the pilots.

An autopilot system is designed in such a way that perform some of the tasks for pilots and helps them and in this way decreasing the load of work from the pilots.

In the present interval of time, Modern autopilot monitors, collision avoidance, sensors, performance, and takes in data from pilots to make the modifications.

With respect to the AI, AI was at first, put forth with the help of John McCarthy in the year of 1956. When John McCarthy held the Dartmouth Conference, which was the first academic conference on the subject.

3.1.3. The Present

Artificial Intelligence is presently being used in several numbers of instances, that are mentioned below:

- Face detection—Airport Security;
- Gmail Spam and Phishing detection;
- Recommendations on YouTube, Flipkart, Amazon, and Netflix;
- Facebook and Twitter feeds;
- Bots in Player Unknown Battel-Ground mobile play games;
- Plagiarism checker for technical papers, thesis, and projects;
- The anticipation of routes in Google Maps;
- Optical Character Recognition in Google Lens;
- Fraud prevention in the web as well as smartphones applications;
- Credit risk assessment;
- Cameras in the Android phones that are driven with the help of artificial intelligence, more specifically portrait mode detecting depth;
- Snapdragon 845 processor, the highest in use in Android phones and Bionic A12, its iOS counterpart supports cutting-edge AI processing;
- Automation in the top 4 private banks in India viz. HDFC, ICICI, Axis & Kotak especially Chat Bot;
- Swiggy generates terabytes of data every week and leverages this data for delivery efficiency and to connect customers to the right restaurant;
- Instagram applies ML to detect the contextual meaning of emoji, which has been steadily replacing slang (for example, a laughing emoticon could replace "lol"). With the help of the algorithmically detecting the sentiments behind the emoticons, Instagram can develop and auto-recommends the emoticons and emoticon hashtags; and
- Uber cabs demand prediction, demand-supply matching, optimal pickup points, Expected Time of Arrival for rides, estimated meal delivery times on Uber Eats, as well as for fraud detection.

3.2. WHAT IS ARTIFICIAL INTELLIGENCE?

The transformative nature of AI in the context of business and society is the main proof. Identical to the internet and the smartphone, AI is an enabler to the technology that will have a far-reaching influence in the life of an individual across all of the area.

3.2.1. What is Artificial Intelligence in the First Place?

In simple terms, the Artificial Intelligence is the branch within the field of computer science that study about how to create or develop the machines which are having the capabilities that are very identical to the human intelligence.

There are several numbers of various maturity levels of the AI which is explained in the below sections that are general AI, super intelligence, and narrow AI.

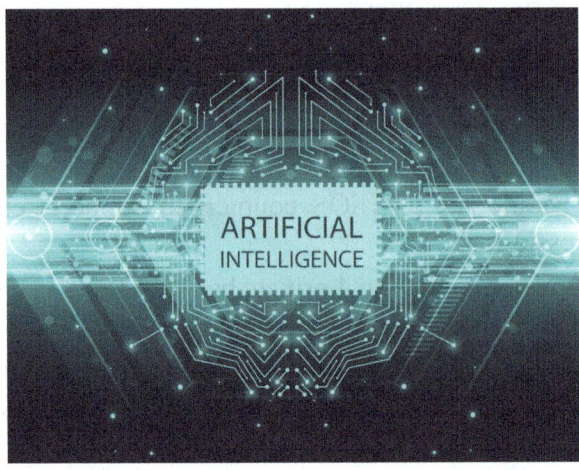

Figure 3.2. The use of artificial intelligence in the first place.

Source: Image by Flickr.

AI is very helpful to digitize the cognitive abilities where the exact rules to follow is a challenging task to explain. A good AI applies case would be the facial recognition. Attempting to utilize the handcrafted knowledge to code all the appropriate rules for facial recognition would be the method, every so often, it is referred to as the first wave of AI. But with the use of mature and reliable technology such as ML and deep learning available, this is not the suggested method to do so.

3.2.2. What Are the Key Drivers of Artificial Intelligence?

Before going deep into the details, let's have a quick look at the main aspects that powers the recent developments in the field of AI:

3.2.2.1. Computing Power

The price-performance of the computing power has increased at a very quick pace with the help of Moore's Law. Quick pace means that the computing speed gets two times and the price slashed by half year over year. In the recent interval of time, ML as one of the most important driving factors of the development of AI, has greatly benefited from the Graphics Processing Units (GPUs).

Graphical processor units (or GPU) play a very important role in order to conduct vectorized numerical functions which are very important with respect to all ML calculations. Tensor Processing Unit (TPU) of Google is the other example where (co)processor is optimized for ML problems. With the help of the improvements in quantum computing it is very likely that this trend will endure for a long time and accelerate as well.

3.2.2.2. Availability of Data

There are an enhanced generation and the availability of the data that is being charged or powered with the help of the elevated application of mobile technology and social media as well. In the interval of the last two years alone, an enormous ninety per cent of the data was generated, all across the world. Huge amounts of data are the prime factor for the success in order to train neutral networks and in this way, accomplish a high precision of their anticipation for the unseen activities.

3.2.2.3. Algorithms

The research society of AI is very active and new developments or improvements are published eventually. The biggest attention in context with the ML, or to be more specific on deep neural networks. There are also several numbers of tools and frameworks that are developed and ready for public use with the help of Facebook (PyTorch), Google (TensorFlow), or Microsoft (Azure) are some examples.

3.2.3. Superintelligence

Several numbers of researchers anticipated and believed that at a certain point of time in the future, the machine will become smarter as compared to human beings. This can take place somewhere between the time period of 2050 and 2100. Several numbers of researchers debated on this and stated that this state would never be accomplished. But with respect to most of the researchers who are familiar with this subject, it is not a matter of whether superintelligence will be accomplished but rather when it will be accomplished.

Superintelligence is the phase where the cognitive abilities of a machine exceed that of human beings. According to Nick Bostrom, it is an organism with an "intellect that greatly surpasses the cognitive performance of the human beings in virtually all areas of the interest."

The beginning of the superintelligence is the basis on the postulation or hypothesis that the speed of the progress is an evolution learning environment that is emerging data very quick pace or exponentially (according to the law of accelerating returns of Ray Kurzweil). This is a difficult task to clutch, and the reason behind that is that most people think rather linearly and try to anticipate the future according to what they already know and have experienced in the past.

It is obvious that, it will take into the direction towards a wrong trajectory. There is a great and fun article written by Tim Urban, who is explaining this journey. It also explains the superintelligence and all its consequences much better as compared to anyone else.

With respect to the best-case scenario, the superintelligence will take into the direction towards the future of abundance and the equity (that is singularity). On the other hand, in the worst-case scenario, it will take into the direction towards the annihilation of manhood, not just due to superintelligence is evil but simply due to the human beings are in its way on accomplishing its objectives (those objectives that human beings are not able to understand any longer from the time they exceed the cognitive abilities of human beings).

3.2.4. General Artificial Intelligence

Before reaching the superintelligence, general AI means that a machine will contain all the cognitive abilities that a human being is having. Again, several numbers of researchers have been debated on the point in time when the human will accomplish the general AI. It could be around the time period of 2014. Because of the law of accelerating returns, the state of the general AI will very soon after the transition into the time period of superintelligence.

Intelligence is extremely connected to the capability of an organism to efficiently mold itself to a better cope with the alterations in the environment. Adaptation also means not only modifying oneself but also modifications of the environment. There are some prime aspects of the general artificial (and human) intelligence that are discussed below:

Figure 3.3. General artificial intelligence.

Source: Image by Flickr.

1. **Learning**: The capability to change the behavior of an individual which is based on past experiences, for example, when encountering new and unseen situations.

2. **Memory**: The encoding, storage, and retrieval of past experiences.

3. **Reasoning and abstraction**: Draw logical assumptions and have the capability to generalize or develop rules which are based on the sample data.

4. **Problem solving**: The capability to systematically come up with possible solutions and develop or create the best answer to the issue.

5. **Divergent thinking**: The capability to generate several numbers of solutions to a given issue.

6. **Convergent thinking**: Narrow down a list of multiple options in

order to derive the best possible answer.

7. **Emotional intelligence**: Recognize and interpret human emotions.

8. **Speed**: All features or aspects that are mentioned above must take place in a reasonable time frame or near real time. Also, they cannot depend on huge amounts of data, for example, to retrain a neural network. In some cases, learning can be based on one single example only.

General-purpose AI will not be accomplished with the help of the present approach or method and tools such as neural networks. In order to accomplish true intelligence, cognitive systems might be a solution.

3.2.5. Narrow Artificial Intelligence

From the time, the researcher did not make much growth in the context of general AI, the main concentration naturally altered to the narrow AI. Narrow AI focuses on particular use cases.

This means that AI (for example, in the present interval of time, a deep neural network is commonly used) is trained for a particular reason. In this way, the AI can only manage activities that it has been trained on.

If a chatbot is trained to answer the customer service request in English for company A, it will not be able to react any request that it will a) receive for a various company or b) are stated in a several numbers of various language or c) be asked for non-related topics such as "Do I need to take an umbrella to work today?."

In difference to general AI, the ability of the narrow AI to learn is very constrained. It can learn within the limits of the specific use case, for example, a narrow AI for speech recognition can be able to enhance the rate of the understanding new languages of the same language it has been trained on.

In order to learn another language, it requires input from human beings, for example, with the help of offering large amounts of labeled data sets for the new language to be learned. Narrow AI cannot dynamically adapt to the new state of affairs coming up with the other solutions, the main trait of general AI.

Some typical narrow Artificial Intelligence use cases:

- Natural language recognition and processing;
- Autonomous driving;
- Visual image recognition and interpretation;
- Intrusion detection in cyber security; and
- Human activity recognition.

3.3. IMPACTS OF ARTIFICIAL INTELLIGENCE IN EVERYDAY LIFE

Whenever there is a discussion about AI, it becomes very easy for people to imagine some out of the world kind of fiction. As soon as AI is discussed, a future is imagined where robots have control over the world along with the human race. But AI is actually a method that makes the human being capable of achieving more simply by collaborating with the software.

Figure 3.4. Impacts of artificial intelligence in everyday life.

Source: Image by Pxhere.

Many researchers have explained AI as giving a human face to the technology. There is an attempt to make turn technology in such a form that is capable of learning from the huge quantities of data that is prevalent in modern times and will even grow more in the future.

AI is making the technology capable of understanding the kind of language that humans speak and then respond in a similar language. With the advancing time, there is a need to make technology self-reliant in such a manner that can view and interpret the world with the same viewpoint as human beings do.

Let us look at some scenarios where Technology can play a huge part in our lives:

- Technology has made the surroundings of human beings easily accessible to them. Imagine if human beings become capable to search and access their surroundings in the same way as the human beings search the web.

By using the existing cameras and advances and including these devices in AI, now it is possible to discover things and people in the real world, in real time. This can prove to be very helpful for human beings, especially when it comes to taking actions in order to improve safety and well-being.

For example, when a dangerous spill or some accident involving chemicals takes place in the chemical plant, the cameras and other technological devices identify that some kind of accident has happened.

Then the devices pass on the information about the accident or spill to the people who are in instant authority and are responsible for the functioning of the zone impacted by spill or accident. This informs the concerned people and thus give them an opportunity to prevent any further damage and set up the things in order.

- Also, technology can prove to be very helpful in making hospitals safer. For instance, there are records maintained in the hospitals that will display the number of patients who have gone under heart surgery. There might be the limited number of patients that might be showing signs of recovery from heart surgery.

So, these patients are advised not to exert much pressure while performing their daily activities. But, when a patient exerts more pressure than the prescribed level, the assigned nurse or any other concerned person is alerted.

Also, the device sending an alert to the nurse locates the nearest wheelchair and the location of the closest wheelchair is identified. This helps in saving the time and helps the nurse in reaching quickly to the patient and keep them safe.

- This technology is very also helpful in a kind of environment such as a construction site where specialized tools that are required by people are spread out, sometimes these are spread out across multiple floors.

Using cameras already in place, this technology can identify a specific tool as well as the closest authorized person who can deliver it saving everyone's time and keeping the workflow moving. The digital and physical worlds have come together to make everybody more secure, safe, and productive with AI.

- With the help of the smartphone, an individual interacts with AI from the obvious features like the features that are built-in such as (Alexa, Siri) to not so obvious ones like the portrait mode that is present in the camera of (Google Pixel 2).

It has been observed that Social Media is becoming a non-separable part of today's life. The feeds that an individual sees in the timeline to the notifications that they receive from these apps, all these things are being curated by AI. This has an impact on most of the decisions that an individual makes.

- Whenever the individuals make use of Google/Apple Maps for navigating or booking an Uber or reservation of a flight ticket, they are using AI. AI is behind many of the products of Google and is considered as a big priority for the company.

It has been observed that the banking and finance industry heavily depends on AI for things which include fraud protection, customer service, chatbots, investment.

- Spam filters that are present in the email inbox and Smart email categorization that individual experience with Gmail are AI-powered. AI neutral networks are useful for E-commerce web application sites.

This helps the E-commerce web application sites to quickly return a wide list of the most appropriate products as well as personalized recommendations on the bottom of item pages, the home page, and email in order to increase the revenue tremendously.

- Brain.fm make use of extensive research and trained experts in order to produce a kind of music written and performed by AI. Receptiviti.ai permits the person to analyze the psychology, personality as well as decision-making style of an individual by inputting a block of text, like a post on the blog or an email.

- Clarke.ai is an AI bot that dials into our conference calls and does all the note taking work for us. It has to be noted that causal chess players regularly make use of AI-powered chess engines in order to analyze their games and practice methods.

AI-based Australian AgTech business firm The Yields in association with Microsoft takes microclimate sensing data and combines it with predictive modeling to help farmers improve their production and reduce their risks.

3.4. ARTIFICIAL INTELLIGENCE PERMEATION AND APPLICATION

AI is defined as the theory and development of computer systems. This is able to perform all the tasks normally, which requires human intelligence, like speech recognition, visual perception, decision-making, and translation between languages.

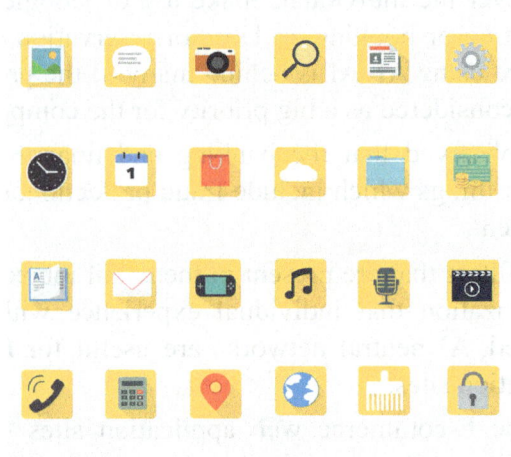

Figure 3.5. Application of the artificial intelligence.

Source: Image by Pixabay.

If the computers can, somehow, solve real-world problems, by improving on their own from past experiences, they would be called "intelligent."

Therefore, the AI systems are more generic (instead of specific). These systems have the ability to "think" and in addition to it, these are more flexible.

AI works its wonders in several numbers of fields. This technological development has made opportunities for people in all spheres of life. Still, it is in its development phase. It has yet to overcome legal, social, ethical, emotional, economic challenges.

The developer's tools and technologies help to exploit the full potential of AI. It comprises of ML algorithms. The kind of AI seen on television like West World, Ex Machina, etc., are way off.

3.4.1. How AI Is Transforming and Revolutionizing Day to Day Things

Facebook's Deep text invades privacy by going through posts, private messages, photos, and videos to understand why they are being posted and accordingly delivers the necessary advertisement. For example, if a person is talking about pizza, then that person may have the advertisement of a pizza popping up.

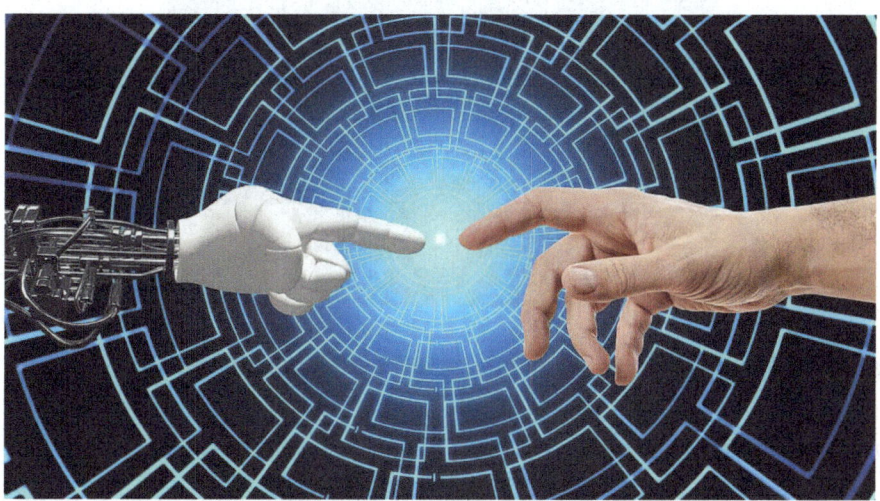

Figure 3.6. Transformation and revolution of Artificial intelligence in day to day life.

Source: Image by Pxfuel

- • Vi, a personal trainer, knows how to deliver the necessary encouragement to maximize running performance by integrating with Google health kit, the weather, and location.
- • Wordsmith can glance into MS Excel or Google Sheets using their free API and can write a detailed analysis of the numbers in seconds.
- • Andrew, a meeting scheduling assistant helps schedule meetings and reminds to check important emails.

3.5. APPLICATIONS OF ARTIFICIAL INTELLIGENCE

3.5.1. Artificial Intelligence in Marketing

Technology has created chances with respect to marketing for a huge market of people. It has altered how people connect to brands, information, services, and technology. Adverts' most required objective which is appropriate at a larger scale is accomplished due to AI. The approaches have made acceptance of brands easier and have offered a solution to the problems of the customers with respect to the marketing platforms.

Figure 3.7. Application of artificial intelligence in marketing.

Source: Image by Pixabay.

It operates on gathering the data, applying the logic to it in terms of profiles, behaviors, patterns, search cycles and taking the important actions with respect to the advertisement and marketing. On the basis of this, it decides which mechanism to apply to attract a loyal customer. It supports the connection between the entrepreneurs, business people, and users. This is a very calculative and precise methodology. It also helps in increasing the target flow of the website.

Some peculiar applications of AI consist of:

1. **Improving the Return on Investment:** The payment process are much faster due to the reason for the high-level image recognition. Better decision-making algorithms are offered on the basis of the gathered data, which is based on the search cycles, patterns, search cycles as well as the behavior which leads to the better database and design. Issues with security with online transactions are solved.

2. **Predicting the Sales**: The trends of digital marketing are put to use only when AI anticipates the future trends of the market. The diverse market can root alteration to the mode of how business works.

3. **Effortless Search Sessions:** This applies the current sessions work well but can be quite confusing. With the help of the AI, working on the digital marketing search engine and the session gets more effective, and the reason behind that is, they track search cycles, patterns, profiles, and behaviors which help in anticipating the behavior of future and decision which could lead to finding accurate and better keyword. It could guide people to the application of the semantic keywords.

4. **elivering to the Right Target Audience Gets Easy:** The importance of any brand elevated when it is delivered to the right people. AI which is based on digital marketing helps to accomplish just that. It also assists in getting the best target audience in context to brand with the help of seeking people on the basis of their demography, interests, and other aspects.

5. **Improvement in Advertising:** Marketing, which is based on AI, in order to create advertisements depending upon the choice and interest of the audience. AI is designed for email marketing, which is consisting of persado, opti-mail, and phrasee. Platforms such as insight pool and cover-social give deep knowledge and work with the impacts of social media marketing.

The predictive analysis which is based on AI is a type of data mining that analyses the historical patterns and anticipates the future patterns which are based on the statistical models. Supply chain optimization by the Wal-Mart, anticipating the trend of price with the help of hopper, and staff retention predictive analysis with the help of IBM.

3.5.2. Artificial Intelligence in Finance

Finance is a part that regulates or manage the lives of each and every individual. The future of money just got more exciting just because of the AI. In order to have smart trading, minimum loss or damage, decreased risks, personalized experience, the prospects are heightened.

E-commerce has acquired its popularity in several numbers of different fields. With the increment in the online fraud cases, in order to fight against this AI has made a breakthrough in the department of finance.

- Banking chatbots such as Plum and Cleo assists in making a customer secure with the help of addressing any queries they have. They also assist in tracking the withdrawal, income, and income of the consumer.

- Algorithmic reading aids in investment funds and stock purchases with the help of gathering and measuring data and searching out the patterns which are then analyzed to develop live trades which are then applied to evaluate profit and loss.

- MasterCard has launched its own tool called 'Decision Intelligence' which keeps tracks of a spending usage of their consumers which assists in order to lessen the cases of fraud.

3.5.3. AI in Healthcare

1. **Managing Medical Records and a Massive Amount of Data:** With the help of AI which assists the researchers to locate the area of concentration for their own research.

It could also assist in seeking the appropriate insights with the help of compilation and evaluating the data.

For instance, the discovery of the disease Amyotrophic Lateral Sclerosis (ALS) was made through a partnership between Barrow Neurological Institute and the AI company IBM Watson Health.

2. **Virtual Nurses:** In AI the conditions of the patient are monitored with the help of nurse bots or virtual nurses such as ly's Molly which helps monitor conditions of patients and examine the treatments while on a doctor visit.

Drug discovery software which is based on AI such as atom wise. Atom wise uses the deep learning techniques which develop the medicine for serious diseases such as ebola with the help of scanning through prevailing medication which could end and cure the disease.

3. **Drug Creation:** Drug creation can anticipate the consequences of the drug treatment and how beneficial a drug is for a person offering a highly personalized method.

Patients who are suffering from cancer are given the same drug and then observed to see the efficiency of that particular drug. It can also help in saving money and time as well, as there is no clinical treatment is needed.

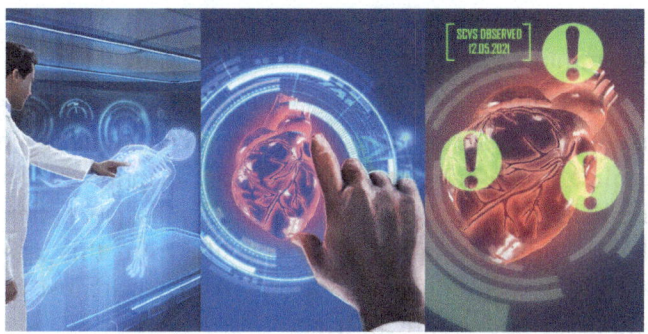

Figure 3.8. Application of artificial intelligence in healthcare.

Source: Image by Flickr.

4. **Personal Life Coach:** As a part of overall care life coaching services are provided. It develops a customized experience for each individual patient and alerts that can be sent to physicians.

5. **Healthcare Bots:** An AI application can assist the patients with their medication, scheduling, requests, billing, and other clinical requirements. They aid all through day and night for primary requests, develop customer service and decrease the administrative costs for the hospitals.

6. **Personal Health Virtual Assistants:** They are very intelligent virtual assistants which when integrated with the healthcare apps can be utilized to offer medication alerts, patient education topic, and human-like interactions to get insight into the current mental state of the patient. They have an incredible influence with respect to monitoring and assisting patients.

7. **Treatment Design:** AI system evaluate the data which contains the reports from the file, external research and clinical expertise of the patient to help and get the individually customized treatment path.

8. **Digital Consultation:** Applications that are based on AI such as Babylon in the United Kingdom offer medical consultation which is based on medical history and medical knowledge of the patient. The database of illnesses is equated to the symptoms of the patient

with the help of speech recognition. The app then guides with the recommended action which is needed to be taken.

9.   **Medication Management:** The National Institutes of Health have created the AiCure app for medication management. In this application, a patient taking their prescription is confirmed with the help of a webcam. This application assists patient to manage their condition. Patients who benefit from this app are patient with serious medical conditions, who go against doctor advice and those under clinical trials

10.   **Precision Medicine:** AI helps in scanning the body, which can spot serious diseases such as cancer and vascular diseases. AI can help in anticipating the health issues that people might face, which are based on their genetics and genomics as they look for mutations and relation to disease based on the information in Deoxyribonucleic acid (DNA).

11.   **Monitoring Health:** Smart bands or smart watches such as Fit bit, Apple, Garmin, and others offer wearable health trackers which observe the activity levels and heart rate as well. User is provided with alerts to exercise more often. This can be shared with doctors and AI systems to analyze the habits and requirements of patients.

12.   **Healthcare System Analysis**: In the Netherlands, near about 97% of   the healthcare invoices are digital. A Dutch company put into use AI to highlight the mistakes in treatments, inefficient workflow and avoiding unnecessary patient hospitalizations.

3.5.4. AI for Automotive Industry

Self-driven cars could decrease the injuries and the harms that took place on the roads. Self-driven cars could also offer a lifestyle choice and people can have more freedom available during driving to do the thing they require.

•   Automotive manufacturing downtime will be reduced with the help of the application of sensors and algorithms to continuously observe the equipment. Huge amount of data can be gathered with the help of the vibration sensors.

It can be used to spot the diagnose problems, anomalies, separate mistakes from background noise and for anticipating if a breakdown will take place.

The Driver-Assist Features is consisting of automatic braking, cyclist alerts, pedestrian, and intelligent cruise control systems and collision-avoiding tasks are being added.

- In the auto insurance industry adjusters applies the Tractable's deep learning systems to make things easier to the triage process after a car accident. In this they use machine-trained estimates for repair costs, enabling agents to hasten a claim past triage and into repair, salvage or appraisal instead of manually scanning pictures.

Equipment uptime could be elevated by near about 20% and the reason behind that is to preventive maintenance, decrease the cost of inspection by near about 25% and lower annual maintenance costs by 10%.

Figure 3.9. Application of artificial intelligence in automation.

Source: Image by Pxhere.

- AI can perceive defects near about 90% more precisely as compared to humans, implement in demand estimating the decrease the excess inventory by up to 50%. As an outcome, in huge cost reduction.

- Tesla has used direct sales model, autonomous driving features, and software performance upgrades to make long drives much safer, with less taxing and convenient as well.

3.5.5. Impact of Internet of Things (IoT) with AI

In the Internet of Things (IoT), any individual can connect any type of device to the internet as well as, to other connected devices. It explains how the connection of sharing influences the environment. IoT does not require a computer or human interaction as it is dependent on internet connectivity.

IoT has devices with inbuilt sensors linked to it which determines the useful information and rejects the pointless information. It can fix schedule variation, check diverse weather conditions, reset alarms. With the help of the sensor-orientation and voice recognition an IoT device can link to a coffee machine and have it ready.

It is anticipated that the IoT will link ten devices to the internet by the year of 2020. Its development relies on how well it is able to regulate the huge amount of data that is being produced. This data is about the consumer, and the data contains his or her behaviors, etc.

With the help of the AI, data can be evaluated or analyzed and these analytics can be utilized to abstract the big data which assists the AI to become more intelligent with the help of learning as it can manage huge and complicated data and to improve the capacity of the AI.

IoT has the power to override the tablets and smartphones. This technology can be utilized in all settings of life from domestic appliances, smart homes to robotics.

3.6. ARTIFICIAL INTELLIGENCE: CAUSE OF UNEM-PLOYMENT

AI has intensely enhanced the world in several numbers of ways, but there are several numbers of concerns with respect to its influence on employment and the workforce.

There are various forecasts that have been made, which states that about hundreds of thousand people getting jobless in the coming several numbers of years, and the reason behind this is the automation and neural networks. There has considerable modification with respect to the education, government sector, business market because of the application of AI.

3.6.1. Sectors Where AI Can Replace Humans

1. **Transportation:** With the development in the automation and ML, people now have the potential to design and develop vehicles that have the capabilities of sensing the environment and move safely with no or little input of manpower.

These types of vehicles are driven by on their own and they do not need any human driver for its movement. With the development in these automated vehicles the need for professionals such as car drivers, pilots, and the pilot would dropdown ata high pace.

2.   **Electronic Commerce:** E-commerce will experience huge transformation because of AI. With the help of the robots, that are navigating the space to gather products or inputs and implement the orders of the customers; to be sent or even delivered to the customers, and that could be done automatically with autonomous cars and drones. In this way, decreasing the requirement for network stores and salespersons.

Figure 3.10. Artificial intelligence: Reason for the unemployment.

Source: Image by Flickr.

3.   **Healthcare:** The time, when AI did not exist, patients always needed a nurse for the means of monitoring the health at a frequent intervals of time and notified the doctor with regards to the health of the patient. But, in the present interval of time, with the help of the devices which are based on AI, attached to the body of patients, in such a way that doctors can monitor the health of the patient at regular intervals of time.

In this way, enabling the doctors to note down observations and take the required decisions with respect to the health of the patient. In this way, there would be no requirement for a nurse to monitor and observe the health of the patients frequently.

4.   **Teaching:** A virtual assistance is a software agent that can be used to do the job for task-based human verbal commands. In the present interval of time, virtual assistants are being designed and developed in such a way that they can teach the scholars just the same a human teacher teaches. In this way, students can

study online with the help of the virtual assistant and it is cost efficient as well. In the coming interval of time, the requirement for teachers and professors would be going to decrease.

3.6.2. Areas Where AI Is Replacing Human Work

1.   **Couriers:** Courier and delivery people are already being replaced by the automated drones and robots within the time frame of several years. The sector of automation would dominate this field. It is forecasted that this field would grow by near about five per cent in the year of 2022.

2.   **Information Technology:** It is anticipated that the automation would increase by near about 12% in the year 2024. In the previous years, for the purpose of testing, manpower is required to tun the code and perform testing. But in the present interval of time, there is no requirement for any human tester as automated testing is performed. In this way, decreasing the requirement of the professional of IT in industry.

3.   **Real Estate:** Real estate plays a very important role in buying and selling of a home. In this age of information, everything is possible on the tips of the finger. Online services such as magic bricks and 99 acres, assist the consumers in seeking out for their properties. In this way, a tech-savvy seller would easily be able to reach the customers without any requirement of any agent.

3.7. DANGERS OF ARTIFICIAL INTELLIGENCE

Now there are several numbers of various AI. But mainly, there are two types of intelligence as all other types of AI are a subdivision of one of these two main types. These two types of AI are mentioned below:

*   Narrow Artificial Intelligence; and
*   General Artificial intelligence

Generally, whenever the topic of Artificial Intelligence comes up, people will only think of only some narrow uses of artificial intelligence. Some of the application of these types of intelligence can be:

*   Playing a chess game;
*   Controlling traffic signals; and
*   Driving a car.

Apart from these, there are other several numbers of the specific applications. In order to implement these all application, the intelligence was given a specific direction in which the system will get the knowledge which is only connected to that particular field or interest.

Hence, it will have no choice of thinking out of that particular type of framework. This type of Intelligence is also known as the "Narrow AI."

The issue that comes up with the "General AI" which is one of the many types of AI. It is full of knowledge about the several numbers of fields and it is also completely open to learning new things on their own. General AI can also learn new things with the help of trial and error method.

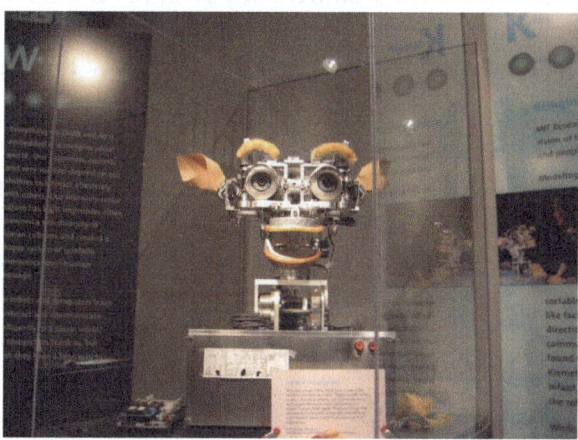

Figure 3.11. Artificial intelligence forecasted as a huge danger.

Source: Image by Wikipedia.

3.8. GENERAL ARTIFICIAL INTELLIGENCE CAN GO AGAINST HUMANS IN THE FOLLOWING WAYS

• Catastrophic failure can take place if the Autonomous control of weapons and arms is given to AI. There is an opportunity that such a huge intelligence can justify to itself that world peace can be attained only by clearing out the human race.

• AI Failure can also take place if it is given a job to do but as an alternative, it chooses a difficult path to do the same job. For instance, if people set a destination for a self-driving car without taking care of any other aspect, it will go to that place with a maximum speed by wreaking havoc over the place it goes.

Hence, the poor designing of AI could take into the direction of these types of havoc. Keeping all these things into consideration, AI should be developed with full accuracy and precision with full care and taking into consideration of all limits should be kept under.

Generally, the algorithm of AI should be made in such a way that:

- It never goes out of control; an
- It gets Knowledge related to only a specific domain.

There is a significant development of AI with a quick pace has been seen from the voice assistant to the self-driving cars. Narrow Intelligence has crossed humans in several numbers of fields, for example, regulating the traffic, playing chess, data management, etc. In the present interval of time, scientists have come to the level of creating a Narrow Intelligence which is a Low-Level AI.

Nevertheless, scientists are still aiming to develop general Intelligence which is a high-level AI. This type of intelligence will have all the governance over the huge data which could even lead to some catastrophic results. General AI has the potential to cross human in every cognitive task possible.

3.9. ADVANTAGES AND DISADVANTAGES OF ARTIFICIAL INTELLIGENCE

AI has turned out to be an integrated part of the modern world. In fact, it has turned out to be one of the emerging technologies in modern times. AI tries to simulate human reasoning in the form of AI systems. JohnMcCarthycoined the term AI in the year 1950.

According to John McCarthy, 'Every aspect of the process of learning or any other aspect of intelligence can, in principle, be so explained in such a precise manner that a machine becomes capable of understanding it and thus will be made to simulate it.'

While explaining the term that he coined, he said, 'In future, an attempt will be made in order to discover how to make the machines use language, form abstractions, and ideas. It will be important to make the machines try to solve the kinds of problems that are reserved only for humans. Moreover, it will be very interesting to watch the machines try to improve themselves, just like human beings do.'

Thus, AI is the capability of any computer program that makes it act like a human and perform human functions such as learn and think. When it comes to classifying the usage of AI, it is simple to recognize the system that applies the AI. Any system or machine can be considered using or a part of AI if it involves a program performing the functions that would normally take human intelligence to plan and execute that function.

3.9.1. Advantages of Artificial Intelligence

There are various advantages of AI. In the modern timers and the future, there are a great number of applications that are capable of completely overhauling or revolutionizing any professional sector or any industry. Some of these advantages are mentioned below:

1. **Reduction in Human Error:** It is evident form the times unknown that human beings commit an error, no matter how small it is. Even according to the famous saying, 'To err is human,' it is a human tendency to commit error before completing any function. On the other hand, computers do not commit any kind of mistakes if the coding and programming have been done properly.

With the introduction of AI, even computer systems are benefitted. Computer systems become capable of making decisions from the previously gathered information. This is done by applying a certain set of algorithms. So, by following this process, the chances of committing errors are reduced. This increases the chances of achieving precision and accuracy with a greater degree of precision is a possibility.

2. **Takes risks instead of Humans:** This is one of the biggest advantages that the human race has achieved with the help of AI. There is a number of risky limitations that human beings cannot go beyond. But, with the help of a Robot developed with the help of AI, human beings can achieve risky things by using the Robot to perform the risky function instead of them.

In fact, there have been a number of operations that would have been not completed if any human being would have tried to perform them. Some of these functions or missions are going to other planets such as Mars, defusing a bomb, exploring the deepest creeks and parts of oceans, performing mining for coal and oil from the core of Earth.

3. **Available 24x7:** According to any studies, it has been found that an average human can work efficiently for almost 4 to 6 hours a day excluding the breaks. The brain of human beings is wired in such a way that it requires some time out for refreshing itself.

Also, the human brain needs a rest in order to get ready to perform its functions the other day. On the other hand, by the use of AI, the machines can be made capable of working 24x7 without any kind of breaks and also the machines do not feel any downgrading emotion like getting bored, whereas the humans feel bored and sleepy after tiresome and continuous working.

4.   **Helping in Repetitive Jobs:** In day-to-day activities, there are various activities that require the repetition of a single activity. Some of these repetitive works are sending a thanking mail, performing the verification of certain documents for errors and similar kinds of tasks. By the use of AI, it has become possible to effectively automate these simple yet time-consuming tasks.

5.   **Digital Assistance:** In the modern world, most of the highly advanced and established organizations take the help of digital assistants so as to interact with users. This helps these organizations in saving the requirement and involvement for human resources. Also, several websites use digital assistants in order to provide and fulfill the requirements of the users. For example, on most of the websites, any user can chat with the digital assistants or 'chatbot' and can have information about the product that they are looking for

6.   **Faster Decisions:** Using AI with a number of other technologies could prove to be very helpful for human beings. By collaborating other technologies with AI, human beings can make machines that will be capable of taking decisions faster than a human and thus implement the plans in a quicker way.

7.   **Daily Applications:** Daily applications such as Apple's Siri, Window's Cortana, Google's OK Google are frequently used in day-to-day human activities. Some of these activities that involve the use of technology are searching a location, taking a selfie, making a phone call, replying to a mail and many more.

8.   **New Inventions:** AI is becoming an important element for powering the majority of the inventions that are taking place in the modern world. These inventions are taking place in almost every domain and will help humans to find the solution for the majority of complex problems.

3.9.2. Disadvantages of Artificial Intelligence

In a similar manner to every other thing in the world, AI has some of the disadvantages that are evident. Some of these disadvantages are:

1. **High Costs of Creation:** As AI is helping everything in evolving, there is a need to update the hardware and software with the passage of every single day. The machines, along with updates, require repairing and service which proves to be very costly. Also, the manufacturing of machines supporting AI needs huge costs as they are very complex machines.

2. **Making Humans Lazy:** With the usage of AI, human beings are becoming lazy with its applications turning most of the basic operations into automated programs. Humans have developed a tendency to get addicted to these inventions which can cause a problem to future generations.

3. **Unemployment:** As AI is being used as the replacement of the majority of the repetitive tasks and by using robots for other works, the requirement of human labor is becoming less day by day. This is turning to be a problem as it is giving rise to a major problem of the employment standards.

4. **No Emotions:** There is no question that machines deliver better output   with higher efficiency as compared to human beings. But machines lack the emotional quotient. Machines cannot develop an emotional bond with human beings. This is very important when it comes to forming a team and executing any task in the form of a team.

5. **Lacking Out of Box Thinking:** Machines are designed in such a way that they perform only those tasks for which they are programmed. But machines are not capable of thinking anything out of the box. This turns out to be the major disadvantage when it comes to using the technology of AI.

3.10. CONCLUSION

With the advancement of technology in modern times, AI has turned out to be one of the most significant parts of technology. The impact of AI is so large that the entire human race is trying to believe its future with the major involvement of machines and robots that would perform major tasks on behalf of human beings.

There are various aspects of AI that are necessary to understand for every human being. AI has turned out to be very helpful in increasing the growth rate as machines are always more efficient than human beings. Also, with the help of AI, machines, and robots are used for tasks that might prove to be very risky for human beings.

But machines lack emotions and emotions are an essential part of the worldly society. Although, attempts are being made to imbibe emotions in machines, it is clear that human beings are not capable of creating another 'robotic human race.'

AI is very beneficial for the future of mankind as it is very beneficial; in industrial processes and could prove to be helpful in saving mankind from unseen disasters.

REFERENCES

1.   Ahmed, H., (2017). Introduction to Artificial Intelligence. [online] https://www.researchgate.net. Available at: https://www.researchgate. net/publication/325581483_Introduction_to_Artificial_Intelligence (accessed on 10 March 2020).

2.   Geeks for Geeks (2018). Artificial Intelligence Permeation and Application—GeeksforGeeks. [online] GeeksforGeeks. Available at: https://www.geeksforgeeks.org/artificial-intelligence-permeation-and-application/ (accessed on 10 March 2020).

3.   Geeks for Geeks, (2018). Artificial Intelligence: Cause of Unemployment—GeeksforGeeks. [online] GeeksforGeeks. Available at: https://www.geeksforgeeks.org/artificial-intelligence-cause-of-unemployment/ (accessed on 10 March 2020).

4.   Geeks for Geeks, (2018). Dangers of Artificial Intelligence – GeeksforGeeks. [online] GeeksforGeeks. Available at: https://www. geeksforgeeks.org/dangers-of-artificial-intelligence/ (accessed on 10 March 2020).

5.   Geeks for Geeks, (2018). Impacts of Artificial Intelligence in Everyday Life—GeeksforGeeks. [online] GeeksforGeeks. Available at: https:// www.geeksforgeeks.org/impacts-of-artificial-intelligence-in-everyday-life/ (accessed on 10 March 2020).

6.   Geeks for Geeks, (2019). Artificial Intelligence | an Introduction— GeeksforGeeks. [online] GeeksforGeeks. Available at: https://www. geeksforgeeks.org/artificial-intelligence-an-introduction/ (accessed on 10 March 2020).

7.   Java T Point (n.d.). Introduction of Artificial Intelligence – Javatpoint. [online] www.javatpoint.com. Available at: https://www.javatpoint. com/introduction-to-artificial-intelligence (accessed on 10 March 2020).

8.   Meruja, S., (2018). Introduction to Artificial Intelligence. [online] Medium. Available at: https://becominghuman.ai/introduction-to-artificial-intelligence-5fba0148ec99 (accessed on 10 March 2020).

9.   Nils, A., (2018). Artificial Intelligence Framework: A Visual Introduction to Machine Learning and AI. [online] Medium. Available at: https://towardsdatascience.com/artificial-intelligence-framework-a-visual-introduction-to-machine-learning-and-ai-d7e36b304f87 (accessed on 10 March 2020).

10. Rajesh, N., (2019). A Brief Introduction to Artificial Intelligence. [online] Medium. Available at: https://medium.com/datadriveninvestor/a-brief-introduction-to-artificial-intelligence-e57eae73e39 (accessed on 10 March 2020).

11. Geeks for Geeks, (2018). What is the Role of Planning in Artificial Intelligence? [online] www.geeksforgeeks.org. Available at: https://www.geeksforgeeks.org/what-is-the-role-of-planning-in-artificial-intelligence/ (accessed on 10 March 2020).

12. Sunil, K., (2019). Advantages and Disadvantages of Artificial Intelligence. [online] Towards Data Science. Available at: https://towardsdatascience.com/advantages-and-disadvantages-of-artificial-intelligence-182a5ef6588c (accessed on 10 March 2020).

Machine Learning: Tools and Techniques

CONTENTS

The chapter of machine learning: tools and techniques explain the basic significance of the tools and the methodologies which are being used in machine learning. This chapter also explains the several types of methods that are used in machine learning such as supervised learning, unsupervised learning, semi-supervised learning, and reinforcement learning.

This chapter mentions the application of tools of machine learning such as why use tools, what are the purpose of using these several tools, how does this serve in delivering the results, and the difference between a good and a great tool of machine learning.

This chapter also provides highlights on the learning techniques which is consisting of multi-task learning, active learning, online learning, transfer learning, and ensemble learning. This chapter emphasizes various the significance of various programming languages such as python, java, C++, and R.This chapter also explains the different frameworks which are used in general machine learning such as NumPy, Scikit-learn, NLTK.

This chapter talks about a part of machine learning that is deep learning, which includes the application of machine learning, untapped opportunities of deep learning, what is required to achieve more with the help of the deep learning. The section also includes the disadvantages of deep learning such as duration of development, amount of data, and computationally expensive training.

4.1. INTRODUCTION

In the last 20 years, Machine Learning has turned out to be one of the major functional areas of information technology (IT). The extensive usage of Machine Learning (ML) in the modern world has made a cerebral yet hidden part of human lives. An increase in the amount of data everyday becomes the cause of making smart data analysis (DA) an important aspect of the IT industry.

It is very hard to analyze the existing data with age-old methods. Thus, it has become very important to adapt new methods to analyze the data. ML is one such method that can be used for analyzing the data even in the future.

According to reports of the several numbers of scientists, they stated that there are several different forms of ML that can appear in several numbers of guises. There a numerous applications that is linked with the application of ML. And also, there are several types of data that are dealt with the help of ML.

After dealing with the data, the final and last step is to present the data in a formalized way. As an alternative, much of th art of ML is to decrease a variety of fairly dissimilar or unlike problems to a set of fairly narrow prototypes. The majority of the science of ML is then to find solutions to these issues and offer a good guarantee for the solutions.

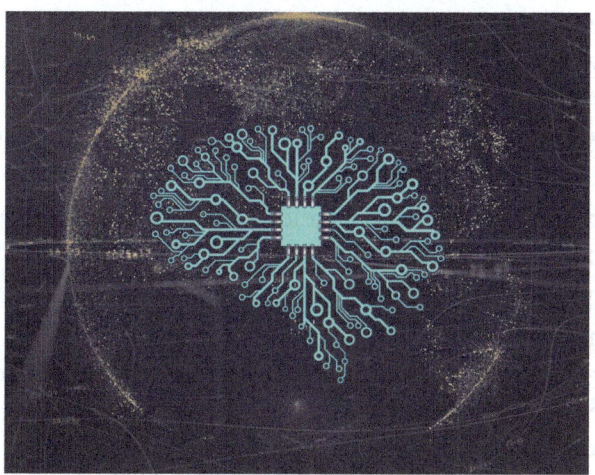

Figure 4.1. The basic significance of machine learning.

Source: Image by Flickr.

Machine learning is a sub-division of artificial intelligence (AI). The main objective of the machine learning commonly is to understand the structure of data and fit the data into the models that can be understood and put up into the use with the help of people.

Nevertheless, ML is a field within the department of computer science, it varies from conventional computational methodologies. With respect to the conventional computing, the algorithms are sets of openly programmed instructions which is being used with the help of the computers, in order to calculate or find the solution of the problem.

On the other hand, the algorithms that are being used in the ML that gives permission to the computer, in order to train on the data inputs and apply statistical analysis to the output values that fall a particular range. Due to this, ML makes it much easier with respect to the computers in developing the models with the help of the sample data in order to automate the decision-making processes which are based on the data inputs.

In the present interval of time, any individual person who uses technology has benefited with the help of ML. Technology such as facial recognition permits platforms like social media to assist the users' tag and share the photos of friends. (OCR) Optical character recognition technology converts the image of text into a movable type.

Recommendation engines, that are driven with the help of ML, recommend what types of movies or television shows to watch, which is based on the preferences of the users. Cars that drive by on their own, that depends on the ML to provide direction, may soon be available to the consumers.

ML is a non-stop developing field. Due to this, there are some major points to keep in mind as an individual is working with ML methodologies or evaluate the influence of the ML processes.

In this chapter, we'll look into the common methods of ML of supervised and unsupervised learning, and the general algorithm methodologies that are being used in the ML, which is consisting of the k-nearest neighbor algorithm, deep learning, and decision tree learning.

This chapter will also explore, what types of programming language are the most used in the context of ML, providing some of the negative as well as positive aspects of each. In addition to this, this chapter will also discuss biases that are continued with the help of ML and consider what can be kept in mind to avoid these biases through the course of development of algorithms.

4.2. MACHINE LEARNING METHODS

With respect to ML, the tasks are commonly categorized into the wide classifications. These classifications are based on how learning is received or how feedback on the learning is given to the system developed.

Two of the most used methodologies or approaches of the ML are supervised learning which trains algorithms which are based on the example input and output data that is labeled by the humans, and unsupervised learning which offers the algorithm with no labeled data in order to permit it to find structure within its input data.

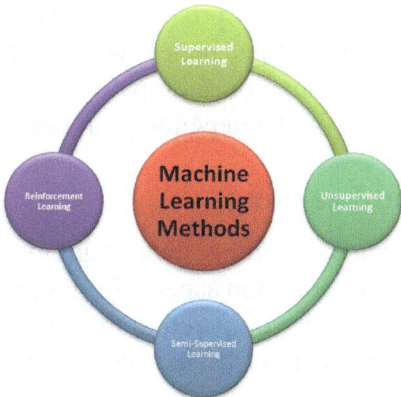

Figure 4.2. Machine learning methods.

4.2.1. Supervised Learning

With respect to the supervised learning, the computer is provided with the example inputs that are labeled with their expected outputs. The main reason behind this method is for the algorithm to be able to "learn" by comparing its actual output with the "taught" outputs to find the error, and alter the model depending on the need. In this way, supervised learning uses patterns to anticipate the label values on the additional unlabeled data.

As an example, with respect to the supervised learning, an algorithm may be fed data with the help of images of sharks labeled as fish and images of the oceans labeled as water. With the help of training on this data, the supervised learning algorithm should have the potential to later identify unlabeled shark images as fish and unlabeled ocean images as water.

The general use case of supervised learning is to apply historical data to anticipate the statistically likely future activities. It can use the historical stock market information to forecast the coming variations or be employed to filter out the spam emails. With respect to the supervised learning, tagged photos of dogs can be used as input data to distinguish the untagged photos of dogs.

In the context of the algorithms that are used in ML, they can apply what has been learned in the past to new data with the help of the labeled examples to anticipate the activities that took place in the coming interval of time. Starting from the analysis of a known training dataset, the learning algorithms create an inferred function to make the anticipation with respect to the output values.

The systems have the potential to provide targets for any new input after the course of satisfactory training. The learning algorithm can also relate its output with the correct, planned or anticipated output and seek out for the errors, in order to alter the model depending on the output.

4.2.2. Unsupervised Learning

In the field of unsupervised learning, data is unlabeled, hence the learning algorithm is left to seek for common aspects from its input data. As unlabeled data are more available as it is compared to the labeled data, the approaches or methodologies with respect to ML, in order to make unsurprised learning easier are particularly valuable. The main objective of the unsupervised learning may be as straightforward as exploring the hidden patterns of it within the datasets. Also, on the other hand, it may have an objective of future learning, which permits the computational machine to explore the representation automatically which is required to categorize the raw data.

Unsupervised learning is generally used for transactional data. An individual can have a large number of a dataset of customers and their purchase, but as a human, an individual will likely not be able to make sense of what are the identical aspects and characteristics can be drawn from the profiles of consumers and their types of purchase as well.

When the data is saved into an unsupervised learning algorithm, it can be determined that the women of particular group of age who buy unscented soaps are likely to be pregnant, and in this way, a marketing campaign that is associated with the pregnancy and the baby products can be targeted to this audience in order to elevate their number of purchases.

Even without giving input of the correct answer, the methods of unsupervised learning can be used to view a complicated sets of data.

Also, these methods of unsupervised learning are very efficient, when it comes to organizing the data, which is more expansive and seemingly unrelated in order, in a potentially meaningful method. Every so often, unsupervised learning is used for variance detection, which is consisting of fraud credit card purchases, and recommender systems that suggest what types of product to buy next.

With respect to the unsupervised learning algorithms, untagged photos of dogs that can be used as input data for the algorithms in order to seek for the identical and classify dog photos together. Unsupervised ML algorithms are generally put into use, when the information used to train is neither classify or labeled.

Unsupervised learning studies how the system can conclude an operation to explain a hidden structure from the labeled data. The system does not find out the correct or appropriate output, but, on the other hand, it discovers the data and can draw conclusions from the datasets to explain the hidden structures from the unlabeled data.

4.2.3. Semi-Supervised Learning

The algorithms of the semi-supervised ML lie somewhere between the supervised and unsupervised learning, from the time being, they use both labeled and unlabeled data for training, which is generally a small amount of the labeled data and a large amount of data.

The systems that apply this type of approach have the potential to enhance the learning accuracy with a considerable amount. Generally, semi-supervised learning is chosen when the gained labeled data necessitates the skills and the appropriate resources, in order to learn from it or train it. On the other hand, acquiring the unlabeled data commonly does not need additional resources.

4.2.4. Reinforcement Learning

Algorithms that supports ML is a learning method that communicates with its environment with the help of producing actions and explores the errors or rewards. Trial and error search and delayed reward are the most appropriate aspects of reinforcement learning.

This methodology or aspect gives permission to the machines as well as software agents to automatically detect the ideal behavior within a particular context to maximize the performance. Simple reward feedback is needed for the agent to learn which action is best, and this is known as the reinforcement signal.

4.3. MACHINE LEARNING TOOLS

Tools play an important role with respect to ML and choosing the right tool can be as significant as working with the best algorithm. In this chapter of ML, you will take a closer look in the context of the tools that are being used in ML. Also, this topic will explain why they are important and the types of tools that an individual could choose from.

4.4. WHY USE TOOLS

ML tools make applied ML with much Faster, Easier, and more fun.

Figure 4.3. Machine learning tools.

Source: Image by Flickr.

1. **Faster**: Tools that are good, can automate each and every step that took place in the process of the applied ML. In simple words, this means that the time from the ideas to the outcomes is greatly shortened. Meanwhile, the substitute to this is, to execute each and every aspect by on your own. This step can take considerably longer as compared to choosing a tool to use off the shelf.

2. **Easier**: An individual can spend his or her time on choosing the good tools as an alternative to the research and execution of the methodologies in order to implement. Working with the substitutes, an individual has to be an expert or professional in every single step of the process to execute the process. This necessitates research, extensive exercise, in order to understand the approaches or methodologies, and a higher level of engineering to make sure that it executes with full efficiency.

3. `**Fun**: With respect to the beginners, there is a lower barrier to get good outcomes. An individual can use the extra time to get better outcomes or work on more projects. The substitute to this is, an individual will spend most of his or her time in developing the tools rather than on getting good consequences.

4.5. TOOLS WITH A PURPOSE

In order to study and have a good hand on the tools of ML, an individual should study on their own sake. They must have a strong reason to do so.

ML tools offer the ability to an individual person that he or she can use to deliver the outcomes in a ML project. A person can apply this as a filter when he or she are trying to choose whether or not to learn about the application of a new tool or new features of the tools.

4.6. HOW DOES THIS SERVE IN DELIVERING RE-SULTS IN A MACHINE LEARNING PROJECT?

Tools of ML are not just the process or series of execution of the algorithms that are used in ML. They can be, but they can also offer the abilities that an individual applies at any point of time through the course of the process of working through ML issues.

4.6.1. Good Versus Great Tools

In order to find the solution to the problems, an individual wants to use the best tools, that he or she is working on. How to do we tell the difference between good and great tools of ML?

1.   **Intuitive Interface**: Great ML tools offer a spontaneous interface onto the sub-tasks of the applied machine process. There is a good mapping and sustainability with respect to the interface for the task.

2.   **Best Practice**: The use of a great ML tool represents the best practice for the configuration, process as well as execution. As an example, the automatic configuration of the algorithms that are being used in the ML and good process built into the structure of the tool.

3.`   **Trusted Resource:** Speaking with respect to the use of great ML tools, are well maintained, they are being updated on a frequent interval of time, and also, they have a community of people around it, who look for the activity around a tool as a sign it is being used.

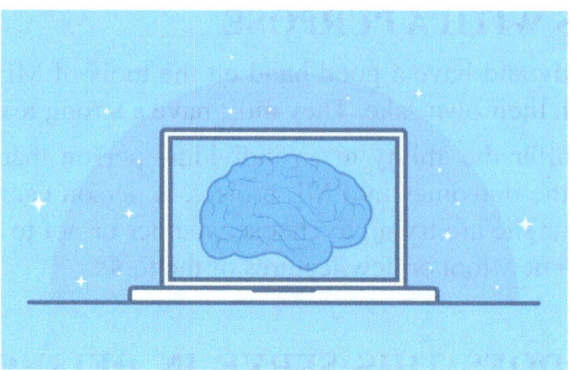

Figure 4.4. Tools of machine learning: how it serves in delivering results.

Source: Image by Pxhere.

4.7. WHEN TO USE MACHINE LEARNING TOOLS

ML tools can save help an individual in saving time and also, it helps in delivering good outcomes across projects. Some cases of when an individual may get the most advantages by the use of ML tools involves:

1. **Getting To start:** When any individual begins to start, ML tools helps them in the process of delivering good results rapidly and in addition to it, it provides confidence to the individual to continue on with their next project.

2. **Day-to-Day:** When an individual wants to get good results to any of the questions quickly, ML tools allow the individual to pay attention to the specifics of the problem in spite of focusing on the depths of the techniques that an individual need to use in order to get an answer.

3. **Project Work:** When an individual is working on a large project, ML tools can help the individual to prototype a solution, figure out all the requirements and it provides an individual with a template for the system that an individual might want to implement.

4.8. PLATFORMS VERSUS LIBRARIES

It has been observed that there are a lot of ML tools. There are enough of tools that a Google search can leave an individual feeling overwhelmed.

One of the important and useful ways to think about ML tools is to separate these tools into Platforms and Libraries. A platform provides all the things to the individual that he or she requires in order to run a project. On the other hand, a library just provides discrete capabilities or parts of what is required by an individual to complete a particular project.

This is not a perfect distinction because some ML platforms are also libraries, or some libraries provide a graphical user interface. Nevertheless, this provides a good point of comparison to differentiate genera case purposes from specific purpose tools.

4.8.1. Machine Learning Platform

A ML platform provides all the capabilities to an individual to complete a ML project from the starting phase to the ending phase. For Example, some DA, modeling, data preparation, and algorithm evaluation and selection.

4.8.2. Features of Machine Learning Platforms

There are several aspects of machine learning platforms. Some of them are:

- They provide capabilities to the person that is required at every single step in a machine learning project;
- The interface might be command line, graphical, programming or all of these or some of the combination;
- They also provide a lose coupling of features. It is required that an individual tie all the pieces together for a particular project; and
- They are designed for general-purpose use and exploration instead of accuracy, speed or scalability.

4.9. EXAMPLES OF MACHINE LEARNING PLAT-FORMS ARE

4.9.1. Machine Learning Library

A ML library provides capabilities to the individual for the completing part or phase of a ML project. For instance, a collection of modeling algorithms may be provided by a library.

Features of ML libraries are:

- They provide a specific capability to the person for one or more than one steps in a machine learning project;
- Usually, the interface is an application programming interface that requires programming; and
- They are designed for a specific use case, environment or problem type.

Some of the examples of ML libraries includes:

- JSAT in Java;
- scikit-learn in Python; and
- Accord Framework in. NET.

4.10. LEARNING TECHNIQUES

Figure 4.5. There are numbers of machine learning techniques that are used in today's world.

Source: Image by Flickr

Learning techniques are the techniques that are used to explain the process that is used to acquire the knowledge of ML. There are a number of techniques through which the knowledge about ML can be acquired. Although there are number of methods to acquire knowledge about ML, there are some common methods. These common methods are multi-task learning, active learning, online learning, transfer learning, and ensemble learning.

4.10.1. Multi-Task Learning

Multi-task learning comes under the classification of supervised learning. This type of technique involves fitting a model on one kind of dataset. This model, then, is used to address and answer a number of related problems.

Multi-task Learning includes finding or creating a model that can be structured or trained on a number of related tasks. This model is devised in such a manner that the performance of the devised model is improved by training across a number of tasks. This kind of training across multiple tasks can deliver better results as compared to being trained on any single task.

Many researchers have provided defining statements for Multi-task Learning. According to these statements, Multi-task learning can be defined as a method to improve generalization by combining the related examples that have come up while executing a number of tasks. These examples are usually the soft constraints that are imposed on the parameters.

As the name suggests, Multi-task learning can prove to be a very useful approach to problem-solving. This method can prove to be most efficient especially when there is an abundance of input data that is labeled for one particular task. This data then, can be shared with some other task that has fewer data labeled for it.

Many learners have used this model of learning when they want to learn multiple related models at the same time. So, for the learners the name of this method validates itself by its usage. This permits the learner to "borrow statistical strength" from a number of tasks that have a huge amount of data and then share and use it with tasks that have a negligible amount of data.

For instance, in a multi-task learning problem, it is common practice to include the same input patterns that can be used for obtaining multiple different outputs or for solving supervised learning problems. In this kid of setup, every output that is obtained may be anticipated or predicted by using a different part of the model. This permits the core of the model to present a generalized pattern across each task for obtaining the same inputs.

In the similar manner that the examples of additional training exert more pressure on the parameters of the model towards values that generalize the model, when the part of a model is shared across a number of tasks, that particular part of the model becomes more constrained towards obtaining good values which usually yields in a better generalization of the model. In this case, it is presumed that the sharing has been done properly.

One of the popular examples of this kind of learning is where the process of same word embedding is used to gain knowledge about a distributed representation of words in text which are then, shared across multiple different natural language processing supervised learning tasks.

4.10.2. Active Learning

Another technique is the technique of Active learning. This is a technique in which the model is able to present a query or question a human user operator in the middle of the learning process. This is usually done in order to find a resolution to the confusion that rises up during the learning process.

In the process of Active learning, the learner collects the training examples in an adaptive or interactive manner. This is done generally by raising a query to an oracle. Mostly this query is to request labels for new points.

Active learning is a kind of supervised learning. This type of learning attempts to attain a similar kind or enhanced performance of so-called "passive" supervised learning. It has been noticed that this type of learning is more effective about the kind of data that is collected or utilized by the model.

The main concept of active learning is that a ML algorithm can be used to attain a greater level of accuracy by using a smaller number of training labels if it is permitted to choose the kind of data from which it learns. An active learner might raise some queries, that might be in the form of unlabeled data instances to be labeled by an oracle (such as, a human annotator).

It is not unreasonable to look at this kind of learning technique as an idea to find the solution of the semi-supervised learning problems, or an alternative model that is used for similar kinds of problems.

In fact, the approach of active learning proves to useful when there is not much data available for study and new data might prove to be very expensive to gather or label.The process of active learning also permits the sampling of the domain to be directed in such a way that it minimizes the number of samples being used and maximizes the efficiency of the model that is being used in the process of learning.

Active learning is generally used in those applications where data gathering or labelling the data prove to be expensive. Computational biology applications are such examples where the data is not available and the process of gathering or labeling the data is very expensive.

4.10.3. Online Learning

Online learning is a process that involves the usage of data that is available and then updating the model directly. The model is updated directly before a prediction is required or after the last observation in the given data was made.

The process of online learning is best suited for the kind of problems where the observations are provided over time. In these kinds of problems, the probability distribution of the observations provided is expected to also change with the passage of time. Therefore, it is expected that the model will change with the same frequency with which the data is made available. This is done in order to capture and harness the changes that come up with the provision of the new data.

Usually the process of ML is performed on offline mode. This means that there is a batch of data that is used to find the answer to the equation. However, if the data is presented in online streaming mode, then the learning has to be done online. This is done in order to update the estimates of the data with the arrival of a new data point. In this kind of learning, there is no point in waiting until "the end" of data. This will not occur as there is a continuous inflow of data.

This approach is also used by algorithms where there may be a greater number of observations than memory can retain. Therefore, online learning is done only over the increasing number of observations, such as a stream of data.

The process of online learning is helpful when the data is changing constantly or rapidly with the passage of time. Also, it proves to be useful for the applications which usually involve a large amount of data that is constantly increasing in size, even if there is a gradual increase in the data points.

It has been noticed that the general approach of online learning is to minimize "regret,." The regret is majorly how well the model has performed in comparison to how well the performance of the model could

One such example of online learning is the so-called stochastic or online gradient descent. This is used to fit into an artificial neural network (ANNs).

The fact that the stochastic gradient descent minimizes generalization error is easiest to see in the online learning case, where examples or mini-batches are drawn from a stream of data.

4.10.4. Transfer Learning

Transfer learning is defined as a kind of learning where a model is trained firstly on one task, then it is trained on some other task or it can also be the case that all of the models is used as the starting point for a related task.

In the case of transfer learning, the learner should perform two or more than two different tasks, but it is assumed that many of the factors that explain the variations in P1 are relevant to the variations that are required to be captured for the purpose of learning P2.

It is considered as a useful approach to problems where there is a task that is related to the major task of interest and the related task consists of a large amount of data.

Also, it has to be noted that it is different from multi-task learning. This is because the tasks are learned sequentially in transfer learning. On the other hand, multi-task learning seeks good performance on all considered tasks by a single model at the same point of time in parallel.

Transfer learning pretrain a deep convolutional net. It has 8 layers of weights on a set of tasks (a subgroup of the 1000 Image Net object categories) and then initializes a network of the same size with the first k layers of the first net.

All the layers that are present in the second network (with the upper layers that are initialized randomly) are then trained together to perform a different set of tasks (another subset of the 1000 Image Net object categories), with less training examples as compared to that for the first set of tasks.

One of the examples is image classification. In this, a predictive model, like an ANN, can be trained on a greater number of general images. On the other hand, when training on a smaller more specific dataset, such as dogs and cats, the weights of the model can be used as a starting point.

The features or aspects that are already learned by the model on the bigger task, like extracting lines and patterns, will be beneficial on the new task that is related.

For example, if there is a presence of considerably more data in the first setting (sampled from P1), then that may be helpful to gain knowledge about representations. These representations are then, useful to present a generalized form of very few examples that are drawn from P2. There are many visual categories that share low-level notions of edges and visual shapes, changes in lighting, the effects of geometric changes, etc.

It has to be noted that transfer learning is especially useful with models that are trained incrementally, and an existing or the current model can be used as a starting point for the purpose of continued training, like deep learning networks.

4.10.5. Ensemble Learning

Ensemble learning is defined as a kind of approach where two or more than two modes are fit on the same data. Also, the predictions from every single model are combined.

The field of ensemble learning provides several numbers of ways of combining the predictions of ensemble members, which includes uniform weighting and weights that are selected on a validation set.

The main purpose of ensemble learning is to attain better performance with the ensemble of models as compared to that of any other kind of individual model. In addition to it, it includes both deciding how to make models that are used in the ensemble and how to combine the predictions from the ensemble members in the best way.

Ensemble learning can be divided into two tasks. This includes:

- Developing a population of base learners from the training data; and
- Then combining this data to create the composite predictor.

Ensemble learning is considered as a useful approach for enhancing the predictive skill on a problem domain. It is also useful in reducing the variance of stochastic learning algorithms, like ANNs.

Some of the examples of popular ensemble learning algorithms consist of:

stacked generalization (also known as stacking), bootstrap aggregation (also known as bagging) and weighted average

It has been observed that bagging, boosting, and stacking have been developed over the past couple of decades. Often, their performance is surprisingly good. There are various ML researchers who have struggled to understand why this is so.

4.11. PROGRAMMING LANGUAGES

At the time of selecting a language to specialize in with ML, an individual might want to consider the skills that are listed on existing job advertisements as well as libraries that are available in a variety of languages that can be used for the processes of ML.

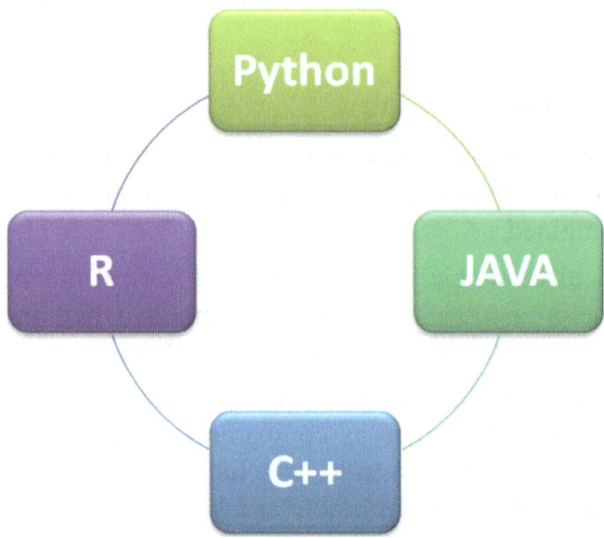

Figure 4.6. Types of programming languages used in machine learning.

From data that is collected from job ads that were posted on the number of job portals in the past few months, it can be assumed that among all the languages, Python is the most sought-for programming language in the field of ML. Python is further followed by other languages such as Java, then R, then C++.

4.11.1. Python

The popularity of Python language maybe because of the increased development of deep learning frameworks that were available for the language in recent times, that includes PyTorch, TensorFlow, and Keras.

Python is the language that has readable syntax and it also has the ability to be used as a scripting language. This language is proved to be powerful and straightforward both for pre-processing the data and functioning directly with the available data.

The scikit-learn ML library is created on top of a number of current Python packages that all the Python developers may already be familiar with, such as SciPy, NumPy, and Matplotlib.

4.11.2. Java

Java is a language that is widely used in enterprise programming, and usually it is used by front-end desktop application developers, those developers who are also working on ML at the enterprise level. Generally, it is not the first choice for those people who are new to programming, who want to acquire knowledge about ML, but it is preferred by those individuals who have a background in Java development to apply to ML.

Java is the language that tends to be used more than Python in terms of ML applications in industry. This is used for network security, which includes cyber-attack and fraud detection use cases.

There are many ML libraries that are used for Java. Deeplearning4j is one such library that is an open-source and distributed deep-learning library. It is written for both Java as well as Scala; MALLET (ML for language toolkit) allows for ML applications on text, which includes, topic modeling, natural language processing document classification, and clustering; and Weka, a collection of ML algorithms that are used for the tasks of data mining.

4.11.3. R

R is an open-source programming language. This language is majorly used for performing statistical computations. R has gained a lot of popularity over recent years. In fact, most of the academic's favor and promote the use of R. When it comes to industrial and production environments, R is not generally used. Yet, the usage of R has gained popularity in industrial applications.

This is majorly just because of the increment in the interest in the field of data science in the industry. Some of the popular packages in the context of ML in R is consisting of CARET (short for Classification and Regression Training). This package is majorly used for creating and building the predictive models.

Another famous package is random Forest. This package is mainly used for classification and regression of the data. Another famous package of R is e1071. This package mainly consists of functions. The functions are mainly used for solving the problems of statistics and probability theory.

4.11.4 C++

One of the most basic and well-known languages is C++. This is the language that is mostly chosen for ML and artificial intelligence (AI) in a game or in some of the major robot applications such as robot locomotion. As more and more languages have been developed in the modern world, yet most of embedded computing hardware developers and electronics engineers favor C++ or C.

This is major because of the level of expertise and proficiency that they have in this language. Along with expertise, these developers and engineers have years of experience of working on this language. That is another reason why C++ is favored more for ML applications.

There are a number of libraries that are present ion the C++ language. Some of these ML libraries that can be used with C++ include the scalable mlpack, Dlib. These libraries usually offer wide-ranging ML algorithms. Another library can be used with C++ is Shark. Shark is an open source library that is modular in nature and can be used for ML applications with C++.

4.12. HUMAN BIASES

Nevertheless, data and computational analysis might force human beings to think that they are receiving objective information, but this is not the case every time. It must be understood that always being based on data does not mean that the outputs that the human being receives from ML are neutral.

The human bias holds an important position in the entire process of ML. Human bias always impacts the way in which the data is gathered and organized. Also, human bias ultimately impacts the algorithms that decide the method in which the ML will interact with the data that is used for applications.

For instance, if the people are providing images for "fish" as data to provide training to an algorithm. Then, if the people choose the images of goldfish, then the computer cannot provide the classification of images of a shark as a fish. This is simply because it would create a bias against sharks as fish, and then sharks would not be considered as fish.

Similarly, there are instances when the historical images of scientists were used in the form of training data. While these images were used as training data, the computer was not able to properly classify the scientists who are also people of Color or women.

As a matter of fact, some of the recent researches that have been reviewed by other scientists show that AI and ML programs usually display biases that are very similar to human biases. Some of these biases that are very evident are race and gender prejudices.

It has been observed that the usage of ML is increasingly leveraged in the business. There have been several uncaught biases that can spread the systemic issues in the system. These kinds of uncaught biases, then, hinder the applications of people from qualifying for loans, from being shown ads for high-paying job opportunities, or from receiving the same-day delivery options.It is a well-known fact that human biases can have a strong negative impact on the thinking and reasoning capabilities of other people. Thus, it becomes extremely important to be aware of the negative impact of human bias. Also, it is not enough to be aware of the negative impact of human bias on the others. The main priority must be to eliminate the negative impact of human bias from the area of work of other people.

There are several methods to eliminate the negative impact of human bias. One such method is to work towards ensuring that the people that are working on a project are from a diverse background and do not have any kind of similarity in their thought process. Also, it should be made sure that these people must have a diversified view of the data when they are testing and reviewing it.

Another method is calling in the regulatory third parties. These third parties usually monitor and audit algorithms. Also, these third parties are responsible for creating alternative systems that are capable of detecting undetected biases, and ethics reviews as part of data science project planning.

The negative impact of human bias in this field can be tackled by raising awareness about biases, becoming more thoughtful with the help of unconscious biases while reviewing the data. Also, it is important to form a structure of equity in the ML projects and pipelines so that any kind of bias that is having an impact on the project can be tackled and removed

4.13 FRAMEWORKS FOR GENERAL MACHINE LEARNING

4.13.1. NumPy: An Extension Package for Scientific Computing with Python

As the name mentions, NumPy is an extension package that is used for performing numerical computing with Python. It replaced the NumArray and Numeric packages. NumPy supports the multidimensional arrays (tables) and matrices. ML data is usually represented in the form of multidimensional arrays.

NumPy consists of broadcasting functions as tools that are used for integrating C/C++ and the Fortran code. The functionality of NumPy also consists of the Fourier transform, linear algebra, and random number capabilities.

Most of the data science practitioners can easily utilize NumPy as an efficient package for storing the generic data, which might be multidimensional in nature. With its feature of defining arbitrary data types, NumPy easily and quickly can be easily associated with different kinds of databases.

4.13.2. Scikit-Learn Easy-to-Use Machine Learning Framework for Numerous Industries

Scikit-learn is a kind of open source Python ML library that is built on the top of SciPy (Scientific Python), NumPy, and matplotlib.

Scikit-learn was initially started in the year 2007 by David Cournapeau. David started this as a part of the Google Summer of Code project. In fact, scikit-learn is still maintained and handled by a number of volunteers from various platforms. To date, near about 1,092 people have contributed to the process of maintaining the scikit-learn.

This library is majorly designed for being used in the production area. It is simple along with having a qualitative code and several collaboration options. Also, the performance and extensive documentation of scikit-learn are coded down in plain language. This is one of the main reasons that has made scikit-learn popular among various specialists.

Also, scikit-learn provides users with several well-established algorithms that can be easily used for supervised and unsupervised learning. Some of the data science practitioners have noted that this library mainly focuses on the modeling data but do not have any emphasis on the process of loading, manipulation, and summarization of the modeling data. Seeing these shortcomings, these practitioners recommend the usage of NumPy and pandas for these three features.

4.13.3. NLTK: Python-Based Human Language Data Processing Platform

NLTK, which stands for Natural Language Toolkit, is a kind of platform that is used for the development of Python programs to function in the form that is compatible with the human language.

Aleksander Konduforov is a well-known data science practitioner who prefers using this tool for NLP tasks. Along with Aleksander, many other data science practitioners have mentioned that NLTK could prove to be a standard library in the language of Python for the tasks of text processing. This library has many features that could prove to be useful while executing ML.

For instance, several kinds of text, sentences, and words processing, part of speech (POF) tagging, sentence structure analysis, named entity recognition, text classification, sentiment analysis, and many other tasks are performed in ML. This library does not incur any cost for its usage. Also, this library has most of the features that provide leverage in the functionality and helps in solving a majority of the tasks in ML.

4.14. DEEP LEARNING: A PART OF MACHINE LEARNING

Deep Learning makes use of ANN. Firstly, an individual will look at a few deep learning applications that will provide an idea of its power.

4.14.1. Applications of Deep Learning

Deep Learning has shown huge success in various areas of ML applications.

- Self-driving Cars: The autonomous self-driving cars make use of deep learning techniques. Usually, they adapt to the traffic situations that constantly change and get better and better at driving over a time period.

- Speech Recognition: The other application of Deep Learning that is interesting is speech recognition. It has been observed that in the present times, every people make use of several mobile apps that are capable of recognizing the speech. There are various examples that make use of deep learning techniques, these are Amazon's Alexa, Apple's Siri, Microsoft's Cortena and Google's Assistant.

- Mobile Apps: An individual makes use of various web-based and mobile apps. They use it for organizing the photos. There are various examples such as face ID, Face detection, face tagging, identifying objects in an image, all these make use of deep learning.

4.15. UNTAPPED OPPORTUNITIES OF DEEP LEARN-ING

All the individual begins to explore other domains where ML was not so far used. They begin to explore it after looking at the great success that the deep learning applications have achieved in many areas.

There are various other domains in which deep learning techniques are applied effectively and also, there are several other domains that can be used. Some of the domains are discussed below:

Agriculture is one such domain where all the individuals can make use of deep learning techniques to enhance the crop yield.

Consumer finance is the other area where ML plays an important role. It helps an individual in providing early detection of frauds and analyzing the ability of the customer to pay.

In addition to it, deep learning techniques are also used in the field of medicine. These techniques are used in order to make new drugs and it also helps in providing a personalized prescription to a patient. The possibilities are endless, and an individual must keep a watch as the new concepts and developments pop up regularly.

4.15.1. What Is Required for Achieving More Using Deep Learning?

Supercomputing power is an obligatory requirement in order to use deep learning. There is the requirement of both memory as well as the CPU in order to develop deep learning models. Fortunately, in the present times, people have easy availability of HPC – High Performance Computing.

Because of this reason, the development of the deep learning applications that that is mentioned above became a reality in the present times. In the future too, one can see the applications of deep learning in those untapped areas that were discussed previously.

In the present times, an individual will look at some of the restrictions of deep learning that one must consider before applying it in the ML application.

4.16. DEEP LEARNING -DISADVANTAGES

Some of the major points that an individual is required to consider before the use of deep learning are listed below:

- Duration of Development;
- Amount of Data; and
- Computationally Expensive.

Now, these limitations will be studies in detail.

4.16.1. Duration of Development

At first, an individual is required to define the problem that one wants to solve, make a specification for it, decide all the input features, design a network, deploy or implement it and further check or test the output.

If the output is not as it was expected by the individual, it should be taken as feedback to restructure the network. This is described as an iterative process and this process may need various iterations until the time network is completely trained in order to produce the required outputs.

4.16.2. Amount of Data

Usually, it has been observed that deep learning networks need a greater amount of data for the purpose of training. On the other hand, the traditional ML algorithms can be applied with great success even if there is the presence of just a few thousands of data points. Fortunately, it has been observed that the data abundance is increasing at 40% every year and in addition to it, CPU processing power is increasing at 20% every year.

4.16.3. Computationally Expensive Training

A neural network needs several times more computational power as compared to the one that is needed in running all the traditional ML algorithms. It has been observed that the effective or successful training of deep Neural Networks may need several weeks of time for the purpose of training.

On the other hand, traditional ML algorithms need only a few minutes or hours for the purpose of training. In addition to it, the amount of computational power that is required for training deep neural networks depends heavily on the size of the data and how deep as well as complicated the network is?

4.17. CONCLUSION

ML is one of the most impost parts of technology that is being used in modern times. In fact, ML has turned out to be the most important technology that can help human beings to step in the future. ML is a sub-field of AI that helps in creating several models of data that can be used for training machines and specifically, computer systems.

The models of data are usually created in such an order that can help the machines in learning from the input given to it. There are various types of ML. The major types are Supervised, Unsupervised, Semi-supervised, Reinforcement learning.

Another type of ML is Deep Learning. Deep Learning is one of the most evolving technology in modern times. With the help of this technology, several futuristic machines such as self-driven cars, automated agricultural machines, and self-operating applications can be developed. ML uses certain language to be operated. Some of the famous languages used in ML are Python, Java, C++ and R. In the modern times and the near future, Python and Java are the most chosen languages for ML.

It has been assumed that most of the outputs that are obtained from solving the ML problems are neutral. These outputs must be neutral in the ideal case. The entire process of ML depends on the kind of input that is given to the system. While some kind of ML gets the data in a single batch, there are some types of ML that receive the input data from online streaming.

In both cases, it is important to consider the impact of human bias on the output received. Not only did the human bias have an impact on the output obtained, sometimes there is a negative impact of the human bias on the output received from the operations of ML. Thus, it is very important to remove the impacts of human bias from the process of ML and try to obtain only the unbiased results.

ML has a great scope in the future and is one of the most important and rapidly growing technology. With the help of ML, most of the machines of the present and future can be trained for the tasks that human beings could not perform. Although it is a very important and significant technology, there is a need to ward off all the negative impacts of human bias from ML.

REFERENCES

1. Altex Soft, (2018). Best Machine Learning Tools: Experts' Top Picks. [online] AltexSoft. Available at: https://www.altexsoft.com/blog/datascience/the-best-machine-learning-tools-experts-top-picks/ (accessed on 10 March 2020).

2. Brownlee, J. (2019). 14 Different Types of Learning in Machine Learning. [online] Machine Learning Mastery. Available at: https://machinelearningmastery.com/types-of-learning-in-machine-learning/ (accessed on 10 March 2020).

3. Expert System Team, (2017). What is Machine Learning? A Definition – Expert System. [online] Expert System. Available at: https://expertsystem.com/machine-learning-definition/ (accessed on 10 March 2020).

4. Lisa, T., (2017). An Introduction to Machine Learning | DigitalOcean. [online] Digitalocean.com. Available at: https://www.digitalocean.com/community/tutorials/an-introduction-to-machine-learning (accessed on 10 March 2020).

5. Tutorials Point, (2019). Machine Learning. [ebook] Tutorials Point. Available at: https://www.tutorialspoint.com/machine_learning/machine_learning_tutorial.pdf (accessed on 10 March 2020).

Social Media Analytics

CONTENTS

In the chapter social media analytics, its application and impact are discussed. The chapter explains the need for social media analytics in detail and its importance for business in detail.

The chapter also explains the various challenges of using social media analytics. In addition, it also explains the social media analytic process.

The chapter also explains the various types of social media analytic techniques like opinion mining, topic modeling, etc.

5.1. INTRODUCTION

Social media refers in a broad sense to a conversational, interactive way of creating, disseminating, and sharing information between communities. Unlike conventional and commercial broadcast media that is of industrial scale, social media has broken down the distinctions between authorship and readership, while the process of content consumption and distribution is intrinsically linked to the process of knowledge generation and its exchange.

5.1.1. Application and Impact

The primary force behind the rise of social media has been the Internet and mobile technology, offering technological platforms for information dissemination, content creation, and interactive communication. However, the majority of social media elements like user-generated content or consumer-generated media have been described as Web 2.0's defining features.

A number of web-based applications describe how social media works from the perspective of a tool. For example, weblogs, microblogs, online forums, life streams, wikis, podcasts, social bookmarks, web communities, social networking, and virtual reality center on avatars.

From an application viewpoint, several social media websites are among the most popular: Wikipedia (collective information generation), Myspace, and Facebook (social networking), YouTube (social networking and sharing of multimedia content), Digg, and Delicious (social searching, news ranking and bookmarking), Second Life (virtual reality) and Twitter (social networking and microblogging) are several of these.

Since social media is already a crucial part of the information environment, and because social media platforms and applications are gaining large-scale acceptance with unparalleled scope for users, customers, voters, companies, governments, and non-profit organizations respectively, interest in social media from all areas of society has grown from both application and from perspectives of research. As both a rich source of information and a business execution forum for product design and innovation customer and stakeholder relationship management and marketing, for-profit companies are tapping into social media.

Social media is an important element of the next generation business intelligence (BI) platform for them. Social media is the perfect medium and knowledge base for politicians, political parties, and governments to assess public opinion on policies and political positions as well as build community support for public office candidates. Authorities in public health can possibly use social media as valuable early indicators about disease outbreaks and provide input on public health policies and initiatives to respond.

Social media provides enormous opportunities to research terrorist group activity for national security and intelligence analysis agencies, including their recruiting and public relations (PRs) strategies and the grounding social and cultural contexts.

Also think tanks and social science and business researchers use social media conceptually as an impartial sensor network and a natural experimental laboratory, providing valuable indicators and assisting in testing hypotheses about social development and interactions as well as their cultural, political, and societal implications.

Social media has become a powerful source of information for many individuals to tackle information-and cognitive-overload concerns, find answers to specific questions, and explore more useful social and economical communication opportunities.

It has also become a platform for them to network and contribute by sharing their expertise and opinions to all kinds of dynamic dialogues. It is fair to say that social media has already penetrated with tremendous effect on a range of applications. Given the ongoing interest and increasing knowledge and meta-information generated by social media, it is expected that new exciting applications will continue to be allowed and many existing ones will be revolutionized.

5.1.2. Social Media Analytics and Intelligence Research

Due to the significant interest from the application's viewpoint and the related specific technological and social science challenges and opportunities, social media research has significantly increased over the past few years.

This research program is multidisciplinary in nature and in all major disciplines has attracted the attention of scientific communities. The research in social media has focused primarily on social media analytics (SMA) and, more recently, social media intelligence from an information technology (IT) perspective.

SMA is involved with designing and evaluating computer tools and frameworks for capturing, tracking, analyzing, summarizing, and visualizing social media data, usually driven by specific target application requirements.

The studies on SMA fulfill several purposes:

- facilitating online communities' conversations and interaction; and

- and extracting useful patterns and intelligence to serve entities that include but are not limited to active contributors in continuing discussions.

SMA research is facing several unique challenges from a technical point of view. First social media provides an abundant set of data or metadata that has not been systematically handled in the literature on data and text mining.

Examples contain tags annotations or identifiers using free-form keywords; user-expressed subjective opinions, observations, perceptions, and perspectives; ratings; user profiles; and social networks both explicit and implicit.

Second, social media applications are a popular example of human-centered computing with its own special emphasis on user-specific social experiences. It is therefore important to re-examine topics such as context-dependent user identification and elicitation needs as well as various types of human computer interaction considerations.

Third, though social media offers a modern approach to addressing the issue of noise and information overload through web-based information processing, problems such as semantic ambiguity, conflicting evidence, lack of structure, inaccuracies, and difficulties in incorporating various types of signals exist in social media.

Fourth, dynamic streams of social media data are rapidly increasing in numbers.The dynamic nature of such data and their sheer size present major challenges to computing in general and, in particular to semantic computing.

Through the use of SMA technologies, solution frameworks and toolsets, social media intelligence aims to extract actionable information from social media in context-rich application environments, build relevant decision-making or decision supporting frameworks, and provide architectural designs and solution frameworks for existing and new applications that can gain from the internet.

Social media intelligence analysis is still at an early stage of development relative to SMA, despite increasing interest from businesses and other groups that might benefit from such studies. From a research perspective, discussions have been conducted on different conceptual aspects of social media knowledge, related technical challenges, and areas of comparison that could theoretically offer valuable tools to help address these challenges.

But there is still a shortage of systematic analysis and practical, well-evaluated results Nonetheless, as a rich, new area of inquiry, social media intelligence offers great potential with considerable practical significance, potentially building on disciplines within AI as well as other fields. There are currently some key challenges facing research on social media intelligence.

First social media intelligence research demands a multidisciplinary study that is highly integrated. While this requirement is often echoed in this growing field, the level of integration continues to be low in existing research.

In many situations, computer science research methodology and research queries play a dominant role, whereas research focusing on methods and issues from other equally relevant disciplines, such as social psychology, media theory, political science, and social sciences, among others, has been dispersed; there have been few highly integrated research programs that consider both computer science and research.

Second, research on social media intelligence includes well-articulated and clearly defined performance measures because much of it needs to be done in application environments with a purpose to facilitate decision-making.

However, quantifying these measures is challenging in a wide range of applications where social media intelligence could be relevant. This issue of estimation makes it particularly difficult to judge the return on investment (ROI) of social media expertise and contributes to difficulties in modeling.

Third, social media intelligence poses a specific class of issues from a pure modeling and decision-making perspective with the need for efficient data-driven, competitive decision-making; complexity and subjective risk analysis; and modeling and optimization through large complex networks.

Researchers will undoubtedly need to develop new theoretical and computational structures and approaches as social media intelligence analysis mature and seek real-time applications.

5.2. SOCIAL MEDIA ANALYTICS

The method of gathering and analyzing data from social media websites and forums to make business decisions is described as Social media analytics or SMA. This process goes outside the normal tracking or a simple retweet review or likes to build the social consumer's in-depth idea.

This is considered the basic basis for allowing a company to: execute targeted commitments such as one-to-one and one-to-many Enhance social collaboration across a variety of business functions such as customer service, marketing, support, etc.

Maximizing the social media customer experience is a good means of understanding consumer choices, intentions, and feelings in real time. SMA most prevalent use is to get to know the customer base on a more emotional level to enable customer service and marketing towards the better target.

During the SMA program, the initial step is to assess which business goals will gain the benefit from the data collected and analyzed. Standard targets include increasing company profits, reducing spending on customer service, collecting input on services and products, and improving public opinion on a business division or particular product. Key performance indicators (KPIs) for carrying out objective data assessment must be defined as soon as the business priorities are determined.

5.2.1. Competitive Advantage

SMA tools enable companies to gain a competitive edge over their rivals by fostering a much better understanding of their brands. It generally includes an awareness of the way consumers make use of certain services or products, the challenges customers face when using these services or products, and the way consumer opinions of a particular company or product are recognized.

5.2.2. Learn from the Customers

In many cases, customers can find effective solutions to some of the organization's problems. For example, if a product is on the market without proper documentation, there is an increased likelihood of mistakes while being in use. Through trial-and-error, some users will solve these issues and then share their results in forums that can help the businesses decide if more documentation is required and what users need to know in real.

5.2.3. Product and Service Improvement

This is SMA's key objective. There are numerous product and service tweets, forums, reviews, and grievances. The vast volume of information includes customer emotions that can be used to measure the interaction of consumers with a specific product or service. Such knowledge can then be used to improve the performance of businesses.

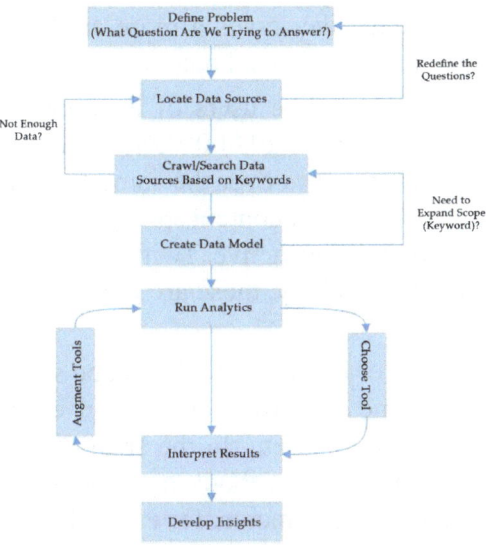

Figure 5.1. Social media analytics.

5.3. THE NEED FOR SOCIAL MEDIA ANALYTICS

PR agencies would monitor customer posts on a company's own website in the early days of social media in an attempt to identify and manage unhappy customers. This is not nearly enough with the increase in the number of social media sites and the amount of use on them.

Just Consider the popularity of social media:

- Social networking is the most popular online practice
- 91% of online adults actively use social media

- Facebook, YouTube, and Twitter are the first, third, and tenth most browsed websites on the Internet, but even these figures do not fully account for the impact of social media. The time users spend is over 20% of their time duration they spent on social media sites.

- Facebook itself has a market penetration rate higher than 12% worldwide; it is 50% in North America. These rates are increasing rapidly, with Facebook alone attracting 170 million new users between the first quarter of 2011 and the first quarter of 2012, a 25% increase. Facebook's smartphone use is increasing much higher at an annual rate of 67%. The volume of information viewed in a single day gives a more startling indication of the enormous influence of social media.

- About one billion active users of Facebook spend collectively on an average of about 20,000 years online each day. YouTube has over 4 billion views over the same twenty-four-hour period, comprising 500 years of content spread across 800 million unique users and 140 million active Twitter users send out over 340 million tweets. Crucially, these are not just passive social media uses.

- The assessment of its videos by YouTube shows that 100 million people take some kind of social action every week by liking, disliking, commenting, etc. In the span of two years, these acts have almost doubled. Facebook now incorporates social actions into its online advertisements, for example by letting users see if their friends have liked or voted to advertise products.

In the same way, hash tags on Twitter and at present other social media platforms have provided users with another convenient way to communicate their likes, dislikes, desires, and concerns, and these pose new opportunities or challenges for companies that want to keep up with these feelings.

5.4. CHALLENGES OF USING SOCIAL MEDIA ANALYTICS

5.4.1. Emergence of Social Networks

Brands report using seven social networks on average, with a participation of 15% of brands in ten or more.

5.4.2. Lack of Consistency throughout the Networks of Metrics

The importance of given social network interaction. Compare Facebook Likes, Views, and Shares and Favorites, Replies, and Retweets of Twitter.

5.4.3. Multiple Social Property Management

Organizations maintain a total of 178 corporate social media assets, as per studies.

5.4.4. Difficulty in Aggregating Properties through Data

Depending on the number of platforms, measures, and resources one needs to manage, even basic reporting will take time.

Data fragmentation and complexity are addressed by these above challenges.

5.4.5. Less Control over Content Subject and Delivery

A portion of the content on social media, such as product news, advertisements, or brand imagery, is not fully controlled by the social media manager.

5.4.6. Susceptibility of External Events Impact

As one example of the cricket World Cup, social media is highly responsive to news and events, resulting in constant fluctuations in results and making comparisons difficult.

5.4.7. Running Campaigns and Initiatives at the Same Time

Often within any given property there is a range of multiple campaigns and programs running simultaneously, so distinguishing their outcomes can be challenging.

5.4.8. Existence of Acts That Are Negative

Because of the variety of experiences, negative effects in the form of dislikes, unfollows, negative comments, etc. have a prominent place in social media. This further complicates the assessment.

The above problems contribute to uncertainty and lack of organization. This is especially problematic for analytics as it is easy to mistake the impact of one variable like the subject for another like the post type without classification. The organization could always be done manually for example in Excel, but this includes an automated solution for content-based sorting and tagging of messages.

5.4.9. Lack of Evergreen Content & Environment for Testing

In social media, there is no such thing as static content that makes testing almost impossible. (Pinned posts and tweets are about the nearest it comes, but for profile/page views only the metrics apply to it.

5.4.10. Extremely Rapid Turnover of Content

The studies have shown that, until they are advertised, the half-life of social posts only lasts about 3 hours. Again, this leaves little room for reviewing and iterating on any content.

5.4.11. No Support of Cohort-Targeted Organic Actions & Results

Targeting or monitoring outcomes was restricted to paid advertising for groups of identified individuals. It makes the organic material of split research unlikely.

5.4.12. Uncontrolled Distribution of Organic Content

Recently, the falling organic reach of Facebook has made waves. This serves as an example of the lack of control by advertisers about whether and the way organic content is being distributed on social media.

5.4.13. Frequent Updates and Improvements to Applications

Social networks, that are typical of new media, frequently update features and functionality, that affect what works and, in some cases, obsolete learning.

The above challenges are the lack of control, especially in terms of organic content and the inability to perform split testing. Although these issues are inherent to the social media world and cannot be easily solved, observations can be derived over time by analyzing unregulated variables through the sheer volume of data. The volume of data generated will be much lower if the problems around the company are tackled first, and the learning can be collected correspondingly quicker.

5.4.14. Inability to Monitor Effect of Engagement

Only for click-through events can organic posts be tracked: interaction activities cannot be traced directly to an off-site operation.

5.4.15. Cross-Effects Between Different Engagement Actions

Shares affect reach that in effect impact likes and comments, but also likes and comments can increase the post's feed rating, raising opportunities for exposure. This jumbling of different influencing actions makes separating their individual impact difficult.

5.4.16. Difficulty of Quantifying Value Per Engagement

It is often difficult even to estimate the value of a share or retweet to the company.

5.4.17. Difficulty in Demonstrating Accurate ROI

The estimation of an accurate ROI for social media is notoriously difficult for all of the above factors. There are case studies, but sales that are directly attributable to social recommendations are usually only a portion of the total effect of revenue.

The issues surrounding the presentation of social media effect will be the hardest to find a solution for, as there is clearly no easy way to connect to influence other marketing activities. Although, there is hope that a possible solution will be found in the future with advancements in marketing effect analysis techniques such as marketing mix or attribution.

5.4.18. Audience and Social Standards Differ Across Networks and Resources

In nature and audience, social networks vary widely, requiring different content strategies.

5.4.19. Difficulty in Obtaining External Expertise

It is often not an option to depend on experienced agencies. Because of social media aspects of branding and customer interaction, only 27% of companies choose to outsource their efforts.

5.4.20. Multiple Roles in Sales Funnel

There has been a time when social media was regarded solely as a function of upper-funnel visibility, but recent developments often make direct response initiatives feasible as well. It makes social media campaigns more nuanced and multi-faceted.

If many of the other problems are addressed, this final set of strategic challenges should be overcome. Once one develops a solid strategy for social analytics and gain information, the full potential of social media as a marketing channel can be applied to strategic goals.

5.5. THE IMPORTANCE OF SOCIAL MEDIA ANALYT-ICS

Social media offers a unique platform for companies to connect with clients and prospects. Although there is a large number of social networks that provide a wide range of tools to provide customer service, demonstrate the way products work, and much more, it is essential to realize that merely having a presence on social media is no guarantee of success.

The social media is very chaotic. It's very competitive. And testing and tracking the results in order to identify the most effective strategies is essential that is why SMA is so important. No data or reviews on the happening on social channels, one is in the dark about what going on and what does not function.

5.6. THE SOCIAL MEDIA ANALYTICS PROCESS

The Analytics of social media involves a process of three stages: capture, understand, and present. The capture stage includes gathering relevant social media data by tracking or listening to different social media outlets, archiving relevant data and extracting relevant information.

This can be done either by an organization itself or by a third-party vendor. Not all collected data will be useful.

The understanding stage chooses relevant data for modeling, eliminates noisy, low-quality data, and uses various advanced data analytics methods to analyze and gain insights from the retained data. The present stage deals with the accurate presentation of Stage 2 results.

The structure is derived from common, widely applied input-process-output models and is consistent with the methodology of Zeng et al., whose monitoring and review activities are subsumed by the comprehension phase; and whose summarization and visualization activities come under the present phase. Among these phases, there is some overlap between them. For example, the stage of learning generates templates that can aid in the stage of capture. Similarly, visual analytics support human judgments that complement the stage of understanding and help in the present stage. In an ongoing, iterative process, these phases are performed rather than purely linear.

If the models struggle to identify useful trends in the comprehension stage, they may be improved by collecting additional data to improve their predictive power. Likewise, if the findings provided are not interesting or have poor predictive power, to adjust the data or change the parameters used in analytics, it may be appropriate to return to the understanding or capture stages. Before it becomes very useful, a system that supports SMA can go through multiple iterations. Before being used by others, data analysts and statisticians help develop and test systems.

5.6.1. Capture

The capture stage helps a company engaged in SMA to recognize interactions related to its activities and interests on social media platforms. This is achieved by using news feeds, APIs, or crawling to gather massive amounts of relevant data through hundreds or thousands of social media outlets.

The capture process includes mainstream channels such as Facebook, Twitter, LinkedIn, YouTube, Pinterest, Google+, Tumblr, Foursquare, etc. as well as smaller, more specialized outlets such as Internet forums, blogs, microblogs, wikis, news sites, photo sharing sites, podcasts, and social bookmarking sites.

Huge amounts of data are archived to meet the different needs of businesses. Numerous pre-processing steps can be performed to prepare a data set for the understanding stage, including data modeling, data or record linkage of data from different sources, stemming, part of speech (POF) tagging, extraction of features, and other syntactic and semantic operations supporting analysis.

In later predictive modeling and interpretation, information about companies, customers, incidents, user comments and reviews, and other entities is also collected. The capture stage needs to balance the need to find information from all areas (inclusivity) with concentrating on the most important and authoritative exclusivity sources in order to assist in a more detailed understanding.

5.6.2. Understand

Once a business has collected the conversations related to its products and operations, it must next assess its meaning and generate metrics useful for decision making. This is the understanding stage. Since the capture stage gathers data from many users and sources, a sizeable portion may be noisy and may need to be removed prior to performing any meaningful analysis.

Simple, rule-based text classifiers or more sophisticated classifiers trained on labeled data may be used for this cleaning function. Assessing meaning from the cleaned data can involve various statistical methods and other techniques derived from text and data mining, natural language processing, machine translation, and network analysis.

This stage provides information about users' sentiments—how they feel about the company and its products and their behaviors including the likelihood of them purchasing in response to an ad campaign, for instance. Many useful metrics and trends about users can be produced in this stage, covering their backgrounds, interests, concerns, and networks of relationships.

Know that the understanding stage is the core of the entire SMA process. The success of this stage will have a significant impact on the information and metrics that are displayed in the present stage, and thus the success of future decisions or actions that might be taken by a firm.

Depending upon the techniques being used and the information being sought, certain analyses may be pre-processed offline while others are computed on-the-fly using data structures optimized for anticipated, ad hoc uses. Humans may participate directly in the understand stage when visual analytics are used to allow them to see various types and representations of data at once or to create visual slices that make patterns more apparent.

5.6.3. Present

The final stage in the cycle of SMA is the present level. The findings of various analyses will be compiled, analyzed, and presented in an easy-to-understand format for users. Different techniques of visualization can be used to display useful information.

The visual dashboard, which aggregates and displays information from different sources, is one of the most commonly used interface designs. Sophisticated visual analytics go beyond pure knowledge show.

They help to make sense of large volumes of information, including trends that are more apparent to people than computers, by promoting personalized views for different users. At this stage data analysts and statisticians might add extra additional support.

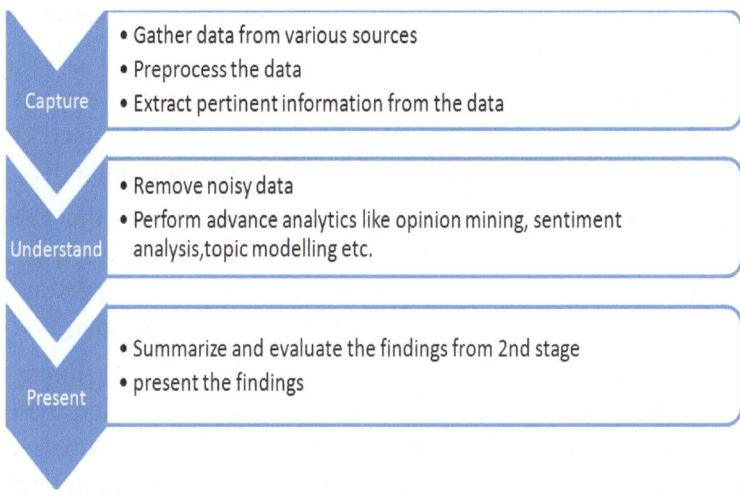

Figure 5.2. Social media analytics process.

5.7. SOCIAL MEDIA ANALYTIC TECHNIQUES

> It should be noted that positive reinforcement refers to not only those stimuli that increase the likelihood of a desirable behaviour but that cause an increase in any behavior (Smith, 2017).

SMA is a growing area that includes a variety of different fields of modeling and analytical techniques. These techniques can support different stages of analytics of social media. Analysis of opinion and evaluating patterns generally help the stage of comprehension.

Topic modeling and analysis of social networks primarily have applications in the process of understanding but can also help the capture and present stages also. Visual analytics encompasses the understanding and current stages.

5.7.1. Opinion Mining

Opinion mining that involves the extraction of thoughts, views or sentiments is a research area that aims to create automatic systems to evaluate human opinion from written text in the natural language.

5.7.2. Sentiment Analysis

The sentiment analysis refers to the use of natural language processing, computer linguistics and text analysis to recognize and extract subjective knowledge in source materials.

Opinion mining or sentiment analysis is an attempt to exploit the vast quantities of online text and news content generated by the user. One of the main features of such material is its textual disorder and wide diversity.

To identify and extract subjective information from the source text, natural language processing, computational linguistics, and text analytics are deployed. The overall goal is to determine a speaker or person's attitude towards some subject or a document's overall contextual polarity.

The sentiment is about mining attitudes, thoughts, feelings—it's not reality, it's subjective experiences. Contrary to popular belief, the objective of the sentiment analysis is to determine the attitude expressed by the writer or speaker toward a document's subject or overall contextual polarity (Mejova, 2009).

Pang and Lee in 2008 stated a detailed information on the basics and approaches of classification and extraction of feelings, including polarity of feelings, degrees of positivity, detection of subjectivity, identification of views, non-factual information, term presence versus frequency, POS (part of speech), syntax, negation, topic-oriented features and word-based features beyond term unigrams.

5.7.2.1. Computational Science Techniques

Digital text automated sentiment analysis uses elements of machine learning (ML) such as latent semantic analysis, vector supporting machines bag-of-words model and semantic orientation (Turney, 2002).

Simply put, the techniques utilize three broad areas: computational statistics defines as computationally intensive statistical methods that include methods of resampling, Monte Carlo methods of the Markov chain, local regression, calculation of kernel density, and analysis of principal components.

ML is a program that acquires and incorporates information from experience, empirical observation, etc. autonomously (Murphy, 2012). Such sub-symbolic structures subdivide further into supervised learning such as Regression Trees, Discriminant Function Analysis, Vector Machines Support.

5.7.2.2. Learning without Supervision Such as SELF-Organizing Maps (SOM), K-Means

ML seeks to solve the issue of having massive amounts of data with several variables and is frequently in use in areas like the pattern recognition like talk, images, financial algorithms like credit scoring, algorithmic trading, energy forecasting (load, price) and biology (tumor detection, drug discovery). (Nuti, 2011)

Complexity science complex simulation models of structures from statistical physics, information theory, and nonlinear dynamics that are difficult to predict. The realm of mathematicians and physicists.

5.7.2.3. Data Mining

exploration of knowledge that derives concealed patterns from huge amounts of data, using advanced differential equations, heuristics, statistical discriminators example hidden Markov models, and artificial intelligence (AI) ML techniques. Examples neural networks, genetic algorithms, and vector machine support.

5.7.2.4. Simulated Modeling

The analysis based on a simulation that tests hypotheses. Simulation is used to try to predict the device dynamics in order to test the validity of the underlying assumption.

5.7.2.4. Stream Processing

The stream processing (Botan, 2010). Analytical applications that consume social media in real time, financial' ticker and data from sensor networks require to process high-volume temporal data with low latency.

These applications need support for rapidly changing data streams for online analysis. Current database management systems (DBMSs), however, have no predefined notion of time and are unable to manage online data in near real time.

This led to the creation of Data Stream Management Systems (DSMSs) that manage transient data streams on-line and process continuous queries on these data streams, processing in main memory without storing the data on disk (Hebrail, 2008). Oracle CEP engine, Stream Base and Microsoft Stream Insight are examples of commercial systems. (Chandramouli, 2010).

5.7.2.5. Sentiment Classification

Sentiment Analysis is categorized in specific subtasks
- Context of Sentiment
- To derive opinion, one needs to know the context of the text, which can vary significantly from specialist review portals or feeds to general forums where views can cover a range of topics (Westerski, 2008).
- Sentiment Level
- Text analysis can be performed at the level of the document, sentence, or attribute.
- Sentiment Subjectivity
- Determining if a given text indicates a viewpoint or is factual that is without a positive or negative opinion being expressed.
- Orientation or polarity of the sentiment
- Making a decision whether a text opinion is positive, neutral or negative.
- Sensitivity strength deciding the power of a definite opinion that in a text whether it is moderate, medium, or powerful.

Perhaps the hardest analysis is to identify the sentiment orientation or polarity and strength of feelings—positively wonderful, elegant, amazing, cool), neutral (fine, ok) and negative (horrible, disgusting, poor, flakey, sucks) because of slang. The text's overall orientation or polarity score is the sum of all found opinion words of all orientation scores. In this simplistic approach, however, there are several potential problems, such as negation. One way of assessing the text's meaning orientation or polarity is pointwise mutual information (PMI), a measure of association used in the theory and statistics of information.

5.7.3. Topic Modeling

It is used to detect dominant themes through large bodies of captured text. The uncovered themes can be used to provide consistent labels to further explore the collection of text or to construct effective navigational interfaces.

Themes discovered by topic modeling can also be used to feed other analytical tasks such as discovering user preferences, detecting emerging themes in forums or posts on social media, or summing up sections or all of a set of text.

The latest developments in topic modeling also make it possible to use these algorithms with Twitter data streaming and other continuous data feeds, making this method an extremely important analytical tool. Topic modeling is based on a range of advanced statistics and techniques for ML.

For example, a number of models define the latest issues through the use of word co-occurrence frequencies in a single communication, or between user topics and communities. Data on the place of words in messages can also be taken into consideration.

Topic modeling is being used to identify dominant trends or subjects from large bodies of collected text from social media platforms. example, Facebook, Twitter, LinkedIn, etc., news articles and behavior of purchase (Bollen et al., 2011; Nelson, Bernstein, and Chi, 2010; Dou, Wang, Chang, and Ribarsky, 2011; Teitler et al., 2008).

For example, the Latent Dirichlet Allocation (LDA), a topic modeling tool (Blei, Ng, and Jordan, 2003), is used to find user interests, recognize critical issues in forums or social media posts (Hennig-Thurau et al., 2010; Kozinets, 1999; Wu, Hofman, Mason, and Watts, 2011). Topic Modeling could also support to identify relationships, like the Who says what to whom (Wu, 2011).

In addition, geographic location labeling (Wang, Xie, and Ma, 2007; Yin, Cao, Han, Zhai, and Huang, 2011), political orientation (Bollen, Mao, and Pepe, 2010; Bollen et al., 2011) and content rumors (Jansen, Zhang, Sobel, and Chowdhury, 2009) could be easily retrieved utilizing topic modeling.

5.7.4. Social Network Analysis

It is used to evaluate a social network graph to learn about its structure, interactions, and theoretical characteristics and to assess the relative importance of various nodes in the network. A graph of a social network consists of nodes or users and related relationships that are shown by edges.

The user acts that directly connect two individuals like as accepting another user as a friend usually identify relationships, although they may be inferred from indirect activities that establish relationships, like voting, tagging, or commenting.

The Analysis of social networks is used to model the dynamics and development of social networks that are network density, positions of new node attachments, etc. that can support to monitor business activities. Social network analysis is the predominant technique used on Twitter or other social media platforms to recognize main influencers in viral marketing campaigns.

It is used to identify sub-communities in a big online community like a discussion forum, making it possible to tailor products and marketing materials more accurately. In predictive modeling, it has strong uses, like conducting marketing campaigns targeted at those most likely to buy a particular product.

The study of the social network uses a variety of related techniques to explain the mathematical structure of graphs. They range from simpler methods like the counting of a node's number of edges or calculating path lengths to more sophisticated methods that calculate eigenvectors that is same the Google's Page Rank algorithm does to classify main nodes in a network. This can be used, for example, to decide who the organization should look at based on its experience, credibility, etc.

The assessment of network structure predates the advent of social media significantly, being developed primarily to analyze static mathematical graphs. The large and constantly changing network structures of today demand new technical approaches, especially when support is sought for real-time decision-making.

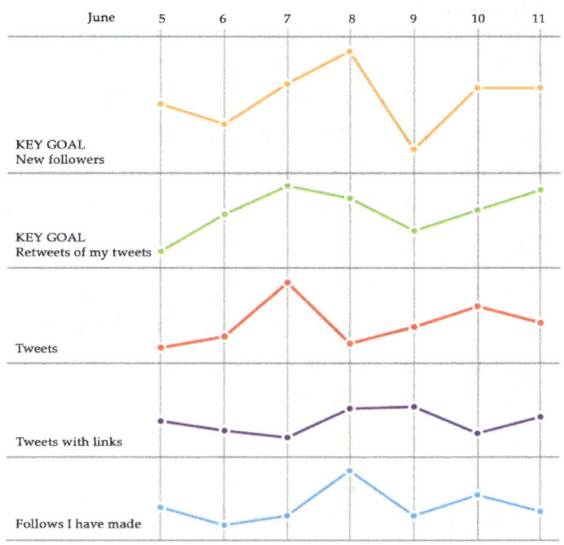

Figure 5.3. Social media analytics of Twitter.

Sources: Image by Flickr.

5.7.5. Predictive Analysis

Predictive analysis refers to the study of historical data and information in order to better understand an entity's future actions such as a person, organization, society, events, etc. For instance, on account of positive recommendations and reviews, Amazon, e-Bay, Pandora, Last.fm, iLike, and many others anticipate customer demand (Miranda et al., 2013; Schifanella et al., 2010).

The predictive analysis can be used for various marketing purposes like churn advertising. (Provost et al., 2009) Taking advantage of prediction tools some scientists also tried to identify the Oscar-winning names prior to the event itself (Bothos, Apostolou, and Mentzas, 2010; Liviu and Mihaela, 2011).

Social post-keywords such as hope, concern, and fear have an important impact on stock indexes (Zhang, Fuehres, and Gloor, 2011). In the same way, emotional keywords such as gold, oil, and currency prices also affect other trends of the financial market (Zhang, Fuehres, and Gloor, 2012).

5.7.6. Visual Analytics

It is the analytical reasoning science supported by interactive visual interfaces. Originally driven by U.S. defense needs, visualization works across various application areas to support synthesis, exploration, discovery, and confirmation of data insights that are generally voluminous and spread throughout different sources. Visual analytics includes a range of activities, from data collection to decision-making that is assisted by data.

While visual analytics are underpinned by many statistical methods like the reduction of high-dimensional data to less and very significant dimensions, the ability of humans to interpret patterns and draw any conclusions are also important factors.

In fact, when there is a flood of information that needs to be accomplished quickly, this combination of machine and human strength is critical, both in making a decision and in being capable of explaining and justify it.

Visual analytics shares a focus on creating economical, intuitive displays with other visualization techniques, but contrary to Tufte's classical work such displays must support real-time decision-making where the stakes can be large. In order to reveal their hidden structure and detail, visual analytics systems must be able to process data.

Computational methods for data reduction, displaying correlations between different sources of data and permitting the user to physically exploit data displays all the visual analytics underlying it.

Visual analytics can be viewed from a more user-perceptual viewpoint as a set of techniques using graphical interfaces to display condensed, heterogeneous information that helps users visually inspect and comprehend the effects of inherent computational processes.

One of the widely used interface designs is a dashboard that depicts various metrics and KPIs in a way that imitates the dashboard design of a car. Using sliders or other types of controls, displays typically allow a user to interrogate the underlying data and perform data transformations. Visual analytics can be of great benefit to both crisis management and the monitoring of breaking news from social media chatter.

The problem for visual analytics is that they are sensitive to increasingly massive and complex data for which a growing number of devices from handheld devices to full-wall display panels will need to be interpreted and displayed more quickly.

5.7.7. Engagement Analysis

Consumer engagement has become another important practice in SMA, since consumers interact more and more with organizations, brands, and products via various social media channels.

It can be used to assess the engagement of consumers involving the brand (Baird and Parassnis, 2011), political engagement of people and voters (Wattal, Schuff, Mandviwalla and Williams, 2010), service development (Claycomb et al., 2001; Graf, 2007) and of brand online communities (Algesheimer, Dholakia, and Herrmann, 2005).

Marketing plays an important role in creating brands in the online environment and uses cost-effective social media.

The financial service provider in New Zealand, for example, ASB, has introduced a virtual branch application that gives users and experts access to the Facebook platform (Talpos, 2011). This application was downloaded 13,000 times in two months, and it was appreciated by 13,600 Facebook users. By October 2015, the figure was up to 136,684.

The Siam Commercial Bank of Thailand also manages customer complaints on social media, such as Facebook and Twitter for the 24*7 week (Miranda et al., 2013). In 2010, HSBC Hong Kong launched its own website in order to allow users to open their own accounts and to read, share, and contact peers and experts (Miranda, 2013).

Additionally, American Express uses social media to make cardholders easier to connect their cards to their Facebook accounts. After connecting customers, customers can view a dashboard with various discount offers from various stores (Heller Baird and Parasnis, 2011).

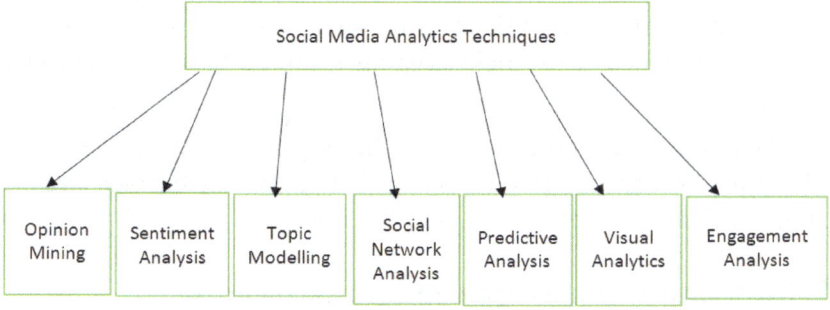

Figure 5.4. Social media analytic techniques.

5.8. THE SOCIAL MEDIA ANALYTICS AND VALUE

There are strong anecdotal evidence of the value SMA offers to organizations, especially in the literature published by practitioners. They know a lot about the benefits that Business Analytics (BA) can achieve, including increased customer profitability, decreased customer defections and higher marketing campaign response rates. However, there are no theoretical, systemic explanations as to the way and why SMA provides the organizations with value and competitive advantage.

5.8.1. SMA Benefits Framework

The social media landscape can be conceptualized as the intersection of social media stakeholders ' activities, including specialist SMA firms and customers, and the level of analysis that investigates these activities.

At the level of management and organization, the focus on social media activities. This comprises the way companies manage and allocate their internal resources when they adopt and use SMA to achieve business goals to create value. The SMA framework is based on RBV, dynamic capacity, IT resources, organizational advantages and motivation for awareness. The SMA framework includes three main concepts: motivation of awareness, SMA resources, and benefits of awareness.

Motivations are the objectives that an organization pursues and guides the organization's subsequent actions. The general motivations that are contained in it are improved customer engagement, increased brand awareness, and increased market value.

SMA tools include successful IT asset combinations and SMA skills that take time to develop and require significant learning and optimization. They also include dynamic capabilities that facilitate organizations to respond by renewing and reconfiguring their existing resource base to environmental turbulence.

The benefits of recognition reflect the degree to which SMA services contribute to the organization's success. Benefits can be measured using indicators of financial, perception, and behavior.

5.8.2. Motivations for Organization Awareness

It is defined as the goals pursued by an organization that guides the organization's subsequent actions. Social media provides the opportunity for organizations on social media platforms to monitor and analyze customer discussions and interactions.

Larson and Watson recognized three motivating consequences in their analysis of the value of social media to organizations: awareness, persuasion, and collaboration. While persuasion and collaboration are important to understand social media's strategic use, they are less relevant to SMA's use.

To create awareness about customer discussions and interactions, they focus on analyzing social media data. Consumer intelligence, marketing, and sales roles within the company are usually associated with awareness.

They describe awareness motivation as the collection and review of social media data to improve organizational knowledge or understanding of customer-empowered environments related issues. Social media data analysis (DA) allows organizations to increase awareness of their brands, products, services, events, marketing campaigns, and overall trends in the market.

SMA's three main objectives of awareness motivation are

- To gain insights into the beliefs and attitudes of customers;
- To understand the impact and efficacy of online marketing campaigns; and
- To discover new concepts for brand image and engagement purposes.

5.8.3. SMA Resources

Awareness motivations drive organizations to develop and renew SMA resources to understand the content, context, and business impact of online posts and conversation (sentiment analysis), to discover valuable customer information (insight mining), and to monitor online user-community relationships (influence analysis, network analysis). SMA resources reinforce mutually IT asset systems (IT infrastructure and SMA tools), SMA capabilities (analytical skills and practices) and vibrant capabilities.

These are IT hardware, software, and networking combinations that provide the basis for shared IT services. They provide a flexible basis for SMA initiatives to enable new applications to be developed to improve business performance.

While traditional information assets involve internal databases and information shared by an organization with business partners, new data assets are required by SMA. Organizations need access to external data streams on social media to quickly identify trends and important issues to drive change in business.

These new IT assets include technologies for storing and analyzing large unstructured databases. example Hadoop. Ultimately, this will make it easier for companies to develop a big data capability.

It is possible to integrate SMA applications with operational IS systems, e.g., CRM. Many BA systems already in place include SMA applications. Examples include predictive analytics for SPSS (Statistical Package for the Social Sciences), and Google Analytics.

Combined with SMA applications, IT solutions provide organizations with text analysis tools, including techniques for sentiment analysis and weak signal analysis techniques for early identification of emerging trends.

5.8.4. SMA Capabilities

These are competency and practices interlocking systems that allow organizations to use SMA applications to perform SMA tasks. For example, an organization needs to allocate suitable IT assets and people with relevant SMA skills to perform a customer insight-mining task.

Organizations need to integrate them into daily business routines in order to interpret and insights gained from that data and take action. Competencies or skills and practices or routines are two key aspects of SMA capabilities within organizations.

SMA skills refer to skills embodied in individuals or teams who are actively managing or carrying out organizational tasks. They are developed by learning and performing related activities repeatedly.

They learn, build skills and develop skills as individuals or teams engage in SMA for a specific purpose. Competencies have two main dimensions, according to Aral and Weill, i.e., IT-related skills, SMA, and business understanding; and management-related skills. Skills related to IT, SMA, and business understanding relate to application development, database management and networking complemented by business skills.

SMA skills include unstructured data management, text mining, processing of natural language, ML, and analysis of social networks. In the newly emerging role of data scientist, SMA skills are prominent.

Management skills include good cooperation with SMA programs and coordination with other company initiatives. Support from senior management drives IT-related projects with adequate funding and clear communication, adding to business value.

Since SMA insights can affect many organizational functions, it may be necessary to adopt a cross-functional approach to SMA. This may include close partnerships with individuals in areas such as marketing and sales, product development and IT. SMA practices serve as a means of performing organizational tasks and as a mechanism for storing and obtaining knowledge on the most effective ways of performing those tasks.

Practices and skills help each other and complement each other. Practices help individuals and teams develop skills with specific ways of working, while skills are needed to effectively execute organizational practices towards specific goals.

SMA practices include IT-related communications and digital transaction practices, Internet architecture usage, evidence-based management, and information management practices. Organizations utilizing digital work practices and proof-based decision-making receive higher performance benefits from SMA as these practices have been embedded in the work routines of their employees.

Personalized SMA dashboards, based on the results of social media,' democratize' information and profoundly integrate contextual insights into the working standards of the enterprise.

Successful utilization of SMA insights depends on three key organizational practices: client management, process management, and performance management. The capacity of a company to consider its customer base e. g. expectations and business dynamics is customer management.

In an SMA context, this is important because it allows companies to enhance their consumer and market intelligence. Process management is the ability of an organization to achieve flexibility, speed, and cost savings by designing and managing key processes effectively.

In an SMA context, process management activities include the application of SMA knowledge into applicable business processes and the implementation of effective indicators and controls.

Performance management is the ability of an organization to design and manage effective output measurement of employees and monitoring systems to support management decision-making and performance communication to relevant stakeholders. If an SMA insight is effectively exercised, the organization can measure its business impact and carry out relevant competitive actions.

5.8.5. Dynamic Capabilities

In volatile environments, the capacity of IT assets and SMA capabilities to deliver organizational benefits can change over time. The RBV has been criticized as being too static in nature, and dynamic technologies have been suggested as a way to refresh and reconfigure IT assets and SMA capabilities to ensure that they continue to deliver advantages and competitive advantage.

As they are not directly involved in the production of goods or the delivery of services, complex capacities are conceptualized because of 'second-order' constructs. Their job is to refresh and reconfigure certain properties and capabilities of the first order to ensure that they remain important, unique, inimitable, and non-replaceable (VRIN). There are two procedures for diverse capabilities: looking for and choosing new innovations and opportunities, and orchestration of assets to refresh and reconfigure assets and capabilities. Searching and choosing routines requires finding ways to leverage SMA to provide enhanced organizational benefits in the sense of SMA tools. It is then possible to prioritize and pick potential SMA opportunities based on their potential business value. The improvisation of assets includes implementing the chosen SMA opportunity by defining and sourcing the necessary IT assets and SMA capabilities.

Dynamic capabilities are organizational routines themselves that have been developed through organizational learning over time.

5.8.6. Awareness Benefits

The three types of awareness benefits achieved with SMA resources are:

- customer-related benefits;
- financial-related benefits; and
- organizational efficiency benefits.

Benefits can be measured using financial (e.g., revenue, costs), perceptual (e.g., customer satisfaction) and behavioral (e.g., SMA insights) measures. Customer-related benefits include a better understanding of customer satisfaction levels, enhanced customer engagement, and a better understanding of customer feelings about products and services. Financial benefits relate to acts resulting from SMA observations that lead to higher sales, lower costs and higher profits.

Financial benefits are often indirectly related to the benefits associated with consumer and organizational performance and may take some time to occur. Organizational efficiency benefits include reduced time-to-market, higher levels of innovation, enhanced flexibility in production and supply chain and enhanced marketing success. They include the use of soft measures and can also have an impact on financial benefits.

5.8.7. Relationships in the SMA Framework

The SMA value framework has two relationships: awareness motivation leads to SMA Resources, and SMA Resources lead to awareness advantages.

The motivation for awareness defines the goals that an organization pursues, which guide the organization's subsequent actions. It is estimated that an organization will develop, outsource, acquire, and renew its SMA resources in order to achieve those goals, depending on the specific awareness motivation.

The described three motivations for awareness: to gain insights into the values and behaviors of customers; to understand the impact and efficacy of online marketing campaigns; and to discover new ideas for brand reputation and interaction purposes. Both of these different motivations will give rise to specific resources for SMA.

For instance, insights into the motivation of customer values and behaviors can lead to the development, outsourcing, acquisition, and renewal of sentiment-related IT assets and SMA capabilities.

SMA resources are the IT assets, SMA capabilities and dynamic capabilities that are relevant to a specific motivation for awareness. It is stated on the basis of RBV that services from SMA contribute to benefits from knowledge. For example, an SMA resource sentiment analysis should lead to better engagement with the customer and brand awareness as sources of benefit from awareness.

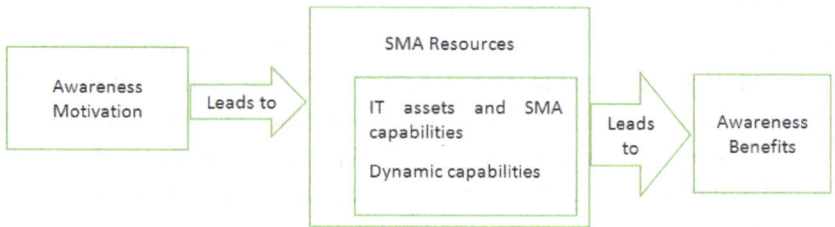

Figure 5.5. Social media analytics benefits framework.

5.9. CONCLUSION

The latest methods used to analyze patterns in social media data to enable informed and insightful decision-making is described as SMA. It gives organizations new methods to create value and benefit from the advantage over the competitors.

In this chapter SMA framework as a means of understanding and the way SMA brings value to organizations has been explained. Through creating concepts from two theories: organizational motivation theory and the resource-based view (RBV) of the organizations this framework has been developed.

The major concern possibly is that companies are increasingly not permitting access to their data to monetize their content. It is due to fear of something wrong will happen that is forcing the firms to take steps in this direction.

It is essential that the researchers should have access to computational environments and mainly big social media data for further development. If not, the big data will become the exclusive domain of big organizations, government agencies its benefits will be denied to the public in the future.

REFERENCES

1. Akter, S., Bhattacharyya, M., Wamba, S., & Aditya, S., (2016). How does social media analytics create value? Journal of Organizational and End User Computing,28(3), 1–9.

2. Batrinca, B., & Treleaven, P., (2015). Social Media Analytics: a Survey of Techniques, Tools and Platforms. [online] Springer. Available at: https://link.springer.com/article/10.1007/s00146-014-0549-4#Sec45 (accessed on 10 March 2020).

3. Bekmamedova, N., & Shanks, G., (2020). Social Media Analytics and Business Value. [online] Ieeexplore. Available at: https://ieeexplore.ieee.org/stamp/stamp.jsp?tp=&arnumber=6759066 (accessed on 10 March 2020).

4. Fan, W., & Gordon, M., (2014). The power of social media analytics. Communications of the ACM,57(6), 74–81.

5. Force, J., (2016). The Importance of Social Media Analytics |. [online] Sysomos. Available at: https://sysomos.com/2016/06/14/the-importance-of-social-media-analytics/ (accessed on 10 March 2020).

6. Techopedia.com. (n.d.). What are Social Media Analytics (SMA)? – Definition from Techopedia. [online] Available at: https://www.techopedia.com/definition/13853/social-media-analytics-sma (accessed on 10 March 2020).

7. Yamaguchi, K., (2014). 20 Reasons Why Social Analytics is a Nightmare—And What to Do About it—Marketing Land. [online] Marketing Land. Available at: https://marketingland.com/20-reasons-social-analytics-nightmare-89676 (accessed on 10 March 2020).

8. Zeng, D., Chen, H., & Lusch, R., (2010). Social Media Analytics and Intelligence. [online] Ieeexplore. Available at: https://ieeexplore.ieee.org/stamp/stamp.jsp?arnumber=5678581 (accessed on 10 March 2020).

Chapter 6

Business Analytics: Its Application Through IoT

CONTENTS

This chapter of Business analytics and its application through the internet of things talk about the basic significance of business analytics and how the internet of things can be integrated into the field of business analytics.

This chapter also explains the various types of analytics, types of tools, and the importance of business analytics. This chapter also provide highlights on the advantages of business analytics such as help in curtailing risks, helps in accurate decision making. helps in accomplishing the efficiency and helps in acquiring valuable insights.

This chapter also explains the applications of the business analytics that most of the organizations are using nowadays. This chapter also includes several numbers of analytical tools such as Sisense, Clear Analytics, Pentaho BI, Micro Strategy, and Qlik View.

This chapter emphasizes the various components of business analytics such as data mining, text mining, forecasting, predictive analysis, optimization, and visualization. This chapter also explains the basic significance of IoT and how to merge the IoT and data analytics to get positive impacts in business.

This chapter also consists of several applications of the IoT that are being used in the present interval of time such as, smart home, wearable's, smart city, smart grids, industrial internet, connected car, etc.

6.1. INTRODUCTION

Business analytics is the process of gathering, organizing, processing, and studying the business data, and also the application of the statistical models and repetitive approaches in order to transform the data into the business insights.

The main objective of the business analytics (BA) is to find out which datasets are useful and how these datasets can be used to find out the solutions to the respective issues and elevate the effectiveness, productivity, and revenue.

BA has a subset that is known as BI. In the context of the BI, BA is commonly executed having the aim of identifying the actionable or useful data.Conventionally, BI is descriptive, and having the focus on the strategies and tools are used to gain, identify, and classify the raw data and prepare the report on the activities or events occurred in past and current interval of time.

On the other hand, BA is more prescriptive, and having the focus on the methodology or approach, with the help of the data can be evaluated, patterns recognized, and models are created to clarify the past activities, create anticipations with respect to the events or activities that will take place in future, and suggest the appropriate action in order to have maximum ideal results.

Quantitative analysis, sophisticated data, and mathematical models are all put into use with the help of business analysts in order to find solutions for data-driven problems.

Also, they can put in use the statistics, computer science, information systems, and operations research to increase their understanding of the data sets which are complicated in nature, and deep learning, artificial intelligence (AI), and neural networks to micro-segment available data and determine the patterns.

This information can then be put in use in order to anticipate the activities that will take place in the future with precision and accuracy, based on the customer's action or trend of the market and to suggest the steps that can attract or influence the consumers towards their required goal.

Figure 6.1. Basic significance of the business analytics.

Source: Image by Wikimedia Commons.

On the other hand, Internet of Things (IoT) remains fundamentally associated together. Data consumed and produced continues to grow with a rate at which it expands forever. This influx of data powers the awareness of the IoT adoption as there will be near about 30.73 billion IoT connected devices by the year 2020.

Internet of Things (IoT) is an affiliation of the various numbers of networks, human resources, devices, and technologies, in order to accomplish a common objective. There are several numbers of applications that uses

IoT, that are being utilized in various sectors and have succeeded in offering enormous advantages or benefits to the consumers.

The generated data with the help of IoT devices, turn out to be of value only if the data gets subjected to the analysis or evaluation, where the data analytics comes into the play. Data Analytics (DA) is defined as a process, which is applied or utilized in order to evaluate small as well as huge data sets, which are having various data properties to abstract the significant outcomes and useful insights.

These extracted conclusions are commonly in the form of patterns, statistics, and trends that helps the business organizations in proactively engaging with the data to execute efficient decision-making processes.

6.2. WHAT IS BUSINESS ANALYTICS?

Business analytics starts with a data set (which is a simple collection of data or a data file) or generally with a database (that is a collection of the data files that is consisting of the information of locations, people, etc.). As the database increase, the data are required to be a store somewhere else.

Figure 6.2. Analytic purpose and tools.

Source: Image by Pixabay

The application of technologies such as computer clouds (hardware and software are used in order to store the data remotely, and for computational functions and retrieval) and the data warehousing (which is a collection of databases that is put into use for reporting and evaluation of data) to store data.

Database storage areas have become so huge that a new terminology was derived, in order to explain these big database storages. Big data describes the group of the data sets that are so complicated and huge, that the software systems are not able to process them (Isson and Harriott, 2013, pp. 57–61).

According to Isson and Harriott (2013, p. 61), the definition of the small data as anything that is not big. Small data explains the little segments of the data or files that provide assistance to the individual businesses to keep track of the consumers. As a means of sorting with the help of the data, in order to find meaningful information, the use of the analytics has found a new reason.

With respect to the Business literature, there are three important terms that every so often associated with each other. These three terms are analytics, BI, and BA. The definition of analytics emphasizes the process that is consisting of application of statistical methodologies and approaches (measures of central tendency, graphs, and so on), information system software (data mining, sorting routines), and operations research methodologies (linear programming) to discover, explore, visualize, and communicate the trends and patterns of the data.

In simple words, analytics transforms the data into useful and meaningful information. Analytics is an older term, which is generally applied to just not the business, but all segments. A typical example of the application of analytics is the measurements of weather, that are being gathered and converted into the statistics, which as an outcome anticipate the patterns of weather.

There are several numbers of types of analytics, and there is a requirement to organize these patterns in order to understand their applications and use them. People will adopt the three classifications of the analytics, that are prescriptive, descriptive, and predictive) that the Institute of Operations Research and Management Sciences (INFORMS) organization (www.informs.org) suggests for grouping the types of analytics.

These three categories of analytics can be viewed as individualistically. For instance, some of the organizations may only use one form (for example, descriptive analytics) in order to provide the information on decisions they cope up with. On the other hand, some may use a combination of analytics types, in order to provide meaningful information that is required to plan and make decisions.

6.2.1. Types of Analytics

1.   **Descriptive:** The use of the simple statistical approaches or methodologies that explains, by what a database or data set is consisting of? For instance, an age bar chart is used to show the retail shoppers for a department store that wants to target the advertising to the customers by age.

2.   **Predictive:** It is an application of the information software, advanced statistical, or operations research methods to find out the predictive variables and create or develop the predictive models, in order to identify the relations and trends, that are not willingly observed in a descriptive analysis.

For instance, several numbers of regression are used to depict the relationship (or lack of relationship) among exercise, age, and weight on diet food sales. Having the idea, that the relationship exists, provide the assistance to describe why one set of the independent variables impacts the dependent variables like business performance.

3.   **Prescriptive:** An application of the management science, decision science, and operations research approaches (applied mathematical techniques) in order to make the best use of the allocable resource. For instance, a department store that has a limited budget to advertise, in order to target the consumers. Linear programming models can be used to distribute the budget to several numbers of advertising media with optimal use of it.

The main reason and the approaches that are used for each of the three kinds of analytics differ, which is explained in Table 6.1. It is these variations that differentiate the analytics from BA.

On the other hand, analytics which is having a concentration on developing meaningful or significant information from the data sources, BA goes the extra step to help the analytics to create the enhancement in measurable business performance.

On the other hand, the process of the analytics can be consisting of anyone out of three types of analytics, the main elements of the BA consist of all three kinds of the analytics are applied sequentially (descriptive, then predictive, then prescriptive).

In this way, the BA can be defined as the process which starts with the collection of data which is related to the business and the process is consisting of consecutive use of major components of the descriptive, predictive, and prescriptive analytic, the results of which help and demonstrates the decision-making and the performance of any organization in business.

According to the Stubbs (2011, p. 11), who have a faith in BA, stated that, BA goes beyond the plain analytics, necessitating a clear application to the business, as an outcome, insight that will be executable, and performance and value measurement to make sure a successful business result.

6.2.2. Analytic Purpose and Tools

Table 6.1. Analytic Purposes and Tools

Type of Analytics	Purpose	Examples of Methodologies
Descriptive	To identify possible trends in large data sets or databases. The purpose is to get a rough picture of what generally the data looks like and what criteria might have potential for identifying trends or future business behavior.	Descriptive statistics, including measures of central tendency (mean, median, mode), measures of dispersion (standard deviation), charts, graphs, sorting methods, frequency distributions, probability distributions, and sampling methods.
Predictive	To build predictive models designed to identify and predict future trends.	Statistical methods like multiple regression and ANOVA. Information system methods like data mining and sorting. Operations research methods like forecasting models.
Prescriptive	To allocate resources optimally to take advantage of predicted trends or future opportunities.	Operations research methodologies like linear programming and decision theory.

According to the definition of Business Intelligence (BI), which states that BI is a group of processes and technologies that transform the data into useable and meaningful information with respect to the businesses. On the other hand, some believe that BI is a broad subject that is consisting of BA, analytics, and information systems (Bartlett, 2013, p.4).

According to some other researchers or analysts, the process of BA is majorly focused on the collection of data sets, storage of datasets, and exploration of large database organizations for information which is significant in order to make decisions and for planning purposes (Negash, 2004).One major operation that is commonly accepted all across the world as a major component of BI is that, it involves storing the huge databases of an organization in data warehouses or in computer cloud storage.

Data warehousing is not an operation of analytics or BA, nevertheless, the data stored in the data warehouse can be used by means of analysis. With respect to the application, BI focused on querying and reporting but it can also include the reported information from an analysis done with the help of BA.

Business analysis is searching for the answers to the questions like what is going on now and where, and also what kind of business actions are required based on the prior experience.

On the other hand, business analysis can answer the questions such as why something is taking place, what kind of new trends may exist, what will take place in the coming interval of time, and what is the best course for the coming interval of time.Business Analytics (BA) is consisting of the similar procedures as in simple or plain analytics but includes the additional desire that the results of the analytic analysis must make a significant or considerable impact on the performance of any business.

Business Analytics (BA) is also consisting of reporting outcomes such as BI, but searching the reasons to describe why the outcomes take place based on the analysis in spite of just reporting and storing the results, just the same as the case of BI.

Analytics, BA, and BI will be mentioned all across this chapter. There is a small review of the characteristics to help in distinguishing these terms, which is presented in Table 6.2.

Table 6.2. Review of the Characteristics of Analytics, Business Analytics, and Business Intelligence

Characteristics	Analytics	Business Analytics (BA)	Business Intelligence (BI)
Business performance planning role	What is happening, and what will be happening?	What is happening now, what will be happening, and what is the best strategy to deal with it?	What is happening now, and what have we done in the past to deal with it?

Use of descriptive analytics as a major component of the analysis	Yes	Yes	Yes
Use of predictive analytics as a major component of the analysis	Yes	Yes	No (only historically)
Use of prescriptive analytics as a major component of the analysis	Yes	Yes	No (only historically)
Use of all three in combination	No	Yes	Yes
Business focus	Maybe	Yes	Yes
Focus on storing and maintaining data	No	No	Yes
Required focus on improving business value and perfor-mance	No	Yes	No

6.2.3. Importance of Business Analytics

- Business analytics is a procedure to make a sound commercial decision. BA influence the operations of the entire organization or business. In this way, BA can help in enhancing and developing the profitability of the organization, helps in boosting its market share and revenue as well, along with that, BA offers a better return on the investments.

- Business analytics helps in acknowledging the availability of primary as well as the secondary data more broadly, which, as an outcome, impacts the operational effectiveness of the various departments that exist in the organizations.

- Business analytics provide assistance in having a competitive edge. In integrates the data which is available with several numbers of various thought models in order to improve the decision making in the business.

- Business analytics helps in converting the available data into meaningful and useable information and also it helps in acquiring the required results and positive results with respect to any organization.

Figure 6.3. Importance of business analytics in modern times.

Source: Image by Pixabay

There is an inactivity of the processes when an organization makes a wrong move. That wrong move could be the reason of slow rate in order to make decisions, just because of the lack of information or proposing an out of the box idea due to the availability of fewer data. This is where the significance of BA comes into play.

6.3. ADVANTAGES OF BUSINESS ANALYTICS

In the present interval of time, each, and every organization is trying to give their best with the help of the BA, when it comes to making decisions for the organization. BA generally consisting of predictive modeling, quantitative, and statistical analysis, data mining, and testing on multi-levels. BA helps in breaking down the past performances in order to make decisions and do the planning for the coming interval of time.

As a matter of fact, it is considered to be very significant, with respect to the economy, more particularly for those who want to get involved in shaping some of the most popular businesses and products.

The significance of the data analytics comes in between the fact that, it has various numbers of layers of applications. There are various numbers of advantages and benefits linked with the BA, but before going deep into business analysis, lets understand it in the much simpler terms first.

Figure 6.4. Several numbers of advantages of business analytics.

6.3.1. Helps in Curtailing Risks

Out of all, this is one of the most important reasons behind the increasing significance of BA. BA helps in making correct and appropriate choices that are based on factors such as changing trends, preferences of the consumers, and performances. This, as an outcome, provides assistance to the organizations in order to avoid the risks for short as well as long term.

6.3.2. Helps in Accurate Decision Making

BA helps in the process of decision making with more precision, because BA helps in understanding the opinions and perspectives of the customer with respect to the organization, company, its product, and its brand as well. With the effectiveness in decision making, the organization can maintain its position on top as compared to their competitors. This is one of the most useful advantages of BA.

6.3.3. Helps in Achieving Efficiency

With the beginning of the BA, the effectiveness with respect to the business has been enhancing regularly. This is one of the most important advantages of BA.

The significance of the BA comes in between the fact that it assists in collecting the data with large volumes at a very quick pace and makes it presentable in a visually attractive way. In this way, the organizations can accomplish particular objectives with a better choice on matters such as where and how to take the business.

6.3.4. Helps in Gaining Valuable Insights

This is one of the best advantages of BA to be considered, and the reason behind this helps in taking care of how the data is presented to the analytics team. There are several numbers of options to choose from such as comprehensive charts and graphs, which provide assistance in order to make decisions more interesting.

6.4. BUSINESS ANALYTICS: APPLICATIONS

In the present interval of time, several numbers of business firms and organizations, all across the world are connecting the power of the data to transform the business scenario.The amount of data that is being generated and gathered is elevating with high pace with every single minute. Nevertheless, simply gathering the data cannot help any organization to grow. It is only with the help of transforming this data into a meaningful quality that organizations can add value to their core foundation.

The main driver to the success comes in between transforming the huge amount of data into meaningful and significant information that can help in terms of scale-up revenue, profitability, and boost the overall effectiveness and productivity of an organization. In this way, the BA courses and jobs are in high demand all across the world.

Figure 6.5. Business analytics and its tools and applications.

Source: Image by Pixabay.

6.4.1. Benefits of Business Analytics

The development of the sector of Information Technology (IT) is very considerable. And also, the IT sector has made the use of BA much appropriate as compared to the previous interval of time. All the meaningful and important data, statistical analysis and computer-based models integrate to perform the process of analysis.

Making suitable and satisfactory decisions with respect to the growth of the company in the coming interval of time and being ready to cope up and address the challenges that an organization might face is the solo objective of the Analytics. It is becoming a huge characteristic of the technical industry.

According to the reports of the Forbes, near about 53% of the tech industry has already accepted this type of culture because of the anticipated benefits and advantages of the BA. It will be the change or modification towards a more important part of the organization and several numbers of issues will be already solved with the help of this before the problem becomes a problem.

The meaningful insights and suggestions provided with the help of the BA tools that permits the organizations as well as various companies, in order to find out the possible ways to automate and optimize the business process. Not only tools of BA provide assistance to companies in order to make data-driven decisions, but they also have several numbers of more clear-cut benefits:

- The use of BA makes tracking and observing the business processes very effective and continuous, in this way, permitting the companies as well as organizations to manage even most complicated business functionalities with comfort.

- The market insights that are provided with the help of BA and BI tools can give a competitive advantage over the competitors as the organization always get updates about the competitors, updates about the potential markets, and also latest trends of consumers. This is highly beneficial with respect to the businesses in a competitive setting.

- Tools that are used in BA (such as predictive modeling, and predictive analytics) can provide more precise and forecasts on time, in the context of present as well as future conditions of the market, on the other hand, BA allows an organization to streamline the marketing strategies for the best possible results.

- With the help of the statistical analysis as well as quantitative analysis, an individual can get a reasonable and meaningful description about why particular types of strategies fail and why others remain successful. In this way, an organization gets a clearer idea of what types of plans they should be concentrating on and what to leave out.

- Tools of BA can effectively evaluate the Key Performance Indicator (KPIs) which can additionally provide assistance to the companies in order to make better analysis on time.

6.5. BUSINESS ANALYTICS TOOLS

The next stop after answering, "What is Analytics" is looking at some tools for the same:

Figure 6.6. Several numbers of tools that are used in business analytics.

6.5.1. Sisense

Sisense is one of the most common tools in the context of analysis in the market. In the year of 2016, Sisense won the Best BI Software Award from the Finances-Online. Sisense is a very helpful tool with respect to simplify the complicated data analyses and making huge data insights feasible and workable with respect to both small and medium-sized organizations.

The key features:

- Sisense's in-chip technology can process data ten times faster than conventional systems.

- It accumulates data from multiple sources with complete accuracy.

6.5.2. Clear Analytics

Clear Analytics is an intelligence tool which is works on excel. Clear Analytics is equipped with several numbers of useful characteristics, for example, version control, sharing capabilities, scheduling of reports, administrative, and governance.

With respect to any individual who is well-versed in the Market Excel, the application of the Clear Analytics will be very easy. Clear analytics is consisting of several numbers of BI oriented characteristics in order to analyze, automate, and visualize all of the suitable data and the information of an organization.

The key features:

- When using Clear Analytics, you don't need a data warehouse as its pre-aggregates data through the Logical Data Warehouse (LDW) approach; and

- Tracing and auditing data are particularly convenient with this tool, so there remains compliance on all levels of the company.

6.5.3. Pentaho BI

Pentaho BI is one of the leading and important tools with respect to the open source BI. Pentaho BI can collect the data from the several numbers of various sources and converts the data into a significant and meaningful insight that can be framed or expressed into a well-articulated campaigns and plans.

To be more precise, Pentaho BI is the ideal tool with respect to those businesses that are looking to increase their profits with the help of faster, better, and precise decision making.

The key features:

- It offers an array of rich navigation features that can enhance data visualization when aided by web-based dashboards.

- The intuitive and interactive analytics of Pentaho BI is equipped with advanced features such as for lasso filtering, zooming, attribute highlight and drill down for improved functioning.

6.5.4. Micro Strategy

Micro Strategy is a very effective tool that permits the organizations to access all the business data from one place nonchalantly. Each and everything are combined into a merged platform so that the business organizations can influence the data to develop significant and compelling platforms.

Micro Strategy uses the powerful dashboards and the data analytics to increase productivity, optimize the revenue, decrease the cost, and anticipate new opportunities, all of which are very important factors with respect to the growth of any organization or company.

The key features:

- It can be used both from mobile devices and desktops.
- Micro Strategy allows you to save data either on-site or in the cloud (powered by Amazon Web services).

6.5.5. Qlik View

QlikView is a platform that is extremely user friendly. QlikView consists of the best of both worlds that is from the tech-based BI tools to the conventional productivity application. QlikView tolls permit the organizations to clutch and process the data in such a way that looks after the innovations.

Whether the organization to improve, provide assistance, and re-engineer to the several numbers of a business process. The application of the QlikView makes sure that an individual come up with efficient as well as exciting solutions for all the requirements of the business.

Key features:

- Qlik View offers a host of customized solutions for sectors such as banking, insurance, etc.
- It is a self-service tool through which businesses can analyze and manipulate data to gain useful insights.

6.5.6. Six Must-Have Features for a Business Analytics Tool

1. **Intuitive Interface:** Intuitive interface should permit the user to perform the analytical functions with the help of an intuitive interface without the help of any type of programming and coding.

2. **Data Blending Capabilities:** From the time, business users get the data with the help of several numbers of various sources, the tool should have enhanced data blending and enrichment capabilities.

3. **Ready to Consume Insights:** In the present interval of time, fast-paced business environment, when a delay of seconds can dispel

several numbers of consumers, the tool should have the potential to deliver ready to customer BI.

4.   **Easy to Share:** It should have easy as well as manageable sharing skills, in order to deliver important and meaningful insights over the network in a multiuser environment.

5.   **Scalable Analytics:** It should be scalable to enable customized analytics and new modules development for the changing business requirements.

6.   **Integration Support:** It should provide assistance to combine or integrate with the other major BI, Business analytics and data visualization tools, in order to enable continuous transportability and compatibility of data.

6.6. COMPONENTS OF BUSINESS ANALYTICS

Business Analytics is the use of statistical tools & technologies to:

- Seek out the patterns in the data in order to perform further analysis. For example, product association.

- Seek out the outliers from the enormous data points. For example, fraud detection.

- Detects the association within the key data variables in order to perform further anticipations. For example, next likely purchase from the Customer.

- Helps in providing meaningful insights as to what is going to happen in the coming interval of time. For example, which of the Customers are leaving?

- Gain a competitive advantage.

So, a more detailed comparison with BI will help to understand better.

Table 6.3. Business Intelligence vs Business Analytics

Business Intelligence	Business Analytics
What does it do?	
Reports on what happened in the past or what is happening now, in current time.	Investigate why it happened & predict what may happen in the future.
How is it achieved?	

• Basic querying and reporting • OLAP cubes, slice, and dice, drill-down • Interactive display options – Dashboards, Scorecards, Charts, graphs, alerts	• Applying statistical and mathematical techniques • Identifying relationships between key data variables • Reveal hidden patterns in data
What does your business gain?	
• Dashboards with "how are we doing" information • Standard reports and preset KPIs • Alert mechanisms when something goes wrong	• Response to "what do we do next?" • Proactive and planned solutions for unknown circumstances • The ability to adapt and respond to changes and challenges

There are six major components or categories in the context of any analytics solution which are explained below.

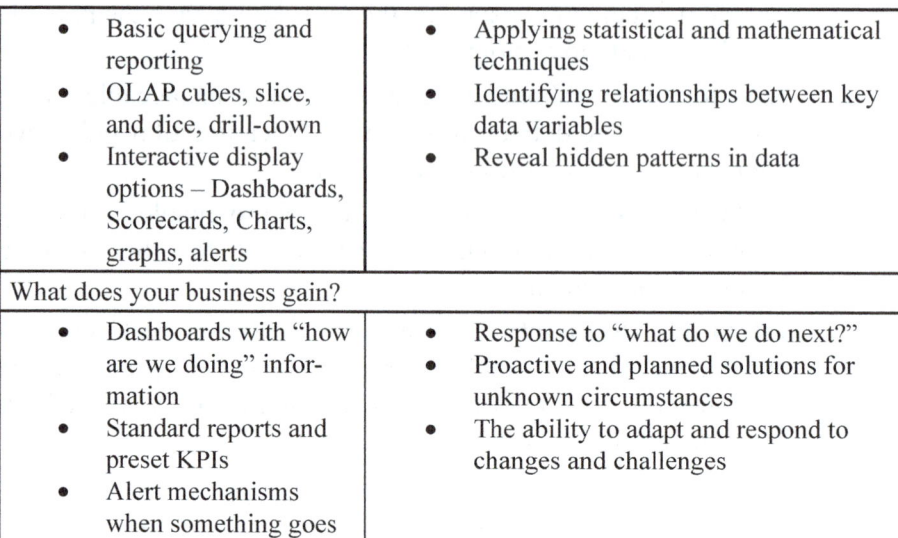

Figure 6.7. Components of the business analytics.

6.6.1. Data Mining

Develops the model with the help of exposing the previously unknown trends and patterns in the huge amount of data. For example, detect retail market basket analysis, and insurance claims frauds.

There are several numbers of statistical techniques with the help of which data mining can be accomplished.

- Classification (when it is understood that, on which variables to classify the data. For example, age, demographics)
- Regression
- Clustering (when it is unclear that, on which types of aspects are needed to classify data)
- Associations and sequencing models

6.6.2. Text Mining

Explore and abstract the significant and useful patterns and relationships from the text collection. For example, understand the sentiments of the customers on social media sites such as Facebook, call center scripts, Twitter, Blogs, etc. which are being utilized in order to enhance the Product or customer service or understand how the competitors are doing.

6.6.3. Forecasting

Evaluate and anticipate the processes that occurred over the interval of time. For example, anticipate the seasonal energy demand with the help of the historical trends, anticipate how many ice creams are needed with respect to the demand.

6.6.4. Predictive Analytics

Develop, regulate, and execute the predictive scoring models. For example, Customer churn and retention, predicting failure, and credit scoring in the shop-floor machinery.

6.6.5. Optimization

Application of identical or repetitive methodologies in order to identify the scenes which will produce the best outcomes. For example, sale price optimization, identifying the optimal Inventory for maximum fulfillment and avoid stock-outs.

6.6.6. Visualization

Improved exploratory data analysis (EDA) and output of modeling, which as an outcome provide results with highly interactive statistical graphics.

6.7. BENEFITS OF DATA-DRIVEN DECISION MAKING WITH BUSINESS ANALYTICS

Small as well as huge companies are using the application of BA in order to make decisions that are derived from the data.

The insights acquired with the help of BA permits these organizations or companies and optimize their business processes. As a matter of fact, data-driven companies that use the BA accomplish a competitive advantage because they have the ability to use the insights in order to:

- Conduct data mining (explore data to seek out for new patterns as well as relationships).
- Complete statistical analysis and quantitative analysis to explain why certain results occur.
- Test previous decisions using A/B testing and multivariate testing.
- Make use of predictive modeling and predictive analytics to forecast future results.

BA also helps in offering support for companies through the course of developing proactive tactical decisions, and BA makes it possible for those organizations to automate decision making in order to support real-time responses.

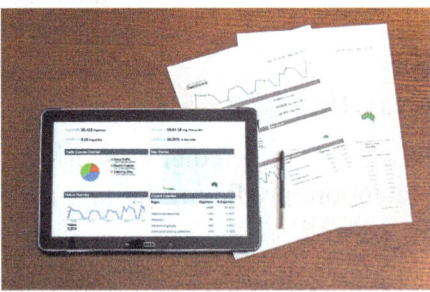

Figure 6.8. Benefits of data-driven making with business analytics.

Source: Image by Pixabay.

6.7.1. The Differences Between Business Intelligence and Business Analytics

Business Analytics (BA) and BA are very identical, nevertheless, they are not exactly the same. BI is consisting of the process of gathering the data with the help of all sources and preparing by means of BA. BI is more of the first step with respect to the organizations to take, at the time when they necessitate the capabilities in order to make data-driven decisions. On the other hand, BA is the analysis if the answers which are provided with the help of BI. Nevertheless, BI answers what took place, BA answers why it took place and whether it will be going to take place.BI consists of automated morning, reporting, scorecards, alerting, ad hoc query, and dashboard. In contrast, BA is consisting of statistical as well as quantitative analysis, predictive modeling, data mining, and multivariate testing.

6.7.2. Challenges with Business Analytics

John Jordan, the researcher at Penn State University, explained the difficult task with respect to BA. There is "a greater potential for the privacy invasion, greater financial exposure in markets which is developing with high pace, greater potential for mistaking noise for true and meaningful suggestions, and a greater risk of spending a lot of money as well as time in order to find the solutions of poorly defined problems or chances." Along with this, there are other difficulties while creating and executing the BA is consisting of:

- Executive Ownership – BA necessitates buy-in from the senior leadership and a clear corporate strategy for combining the predictive models.

- IT Involvement – Technology infrastructure, as well as tools, must have the potential in order to manage the data and BA processes.

- Available Production Data vs. Cleansed Modeling Data – Watch for technology infrastructure that constraint the data is available for historical modeling, and knows the variations between the historical data for model development and real-time data in production.

- Project Management Office (PMO) – The correct project management structure must be in place in order to execute the predictive models and accept an agile approach.

- End-user Involvement and Buy-In-End users should be consisting of adopting BA and have a stake in the predictive model.

- Change Management – Organizations should be prepared for the alterations that BA brings to present business as well as technology operations.

- Explainability vs. the "Perfect Lift" – Balance building precise statistical models with being able to explain the model and how it will produce results.

6.7.3. Business Analytics Best Practices

Accepting and executions of BA is not something an organization can do over the period of the night. But also, if an organization follows some best practices in the context of BA, they will get the levels of insights they are looking for and become more successful and competitive.

There is some list of most significant and necessary best practices with respect to the BA, nevertheless, the organization will require to find out which best practices are most fitting according to their needs, which is mentioned below.

- Know the objective of using BA. Define the business use case and the goal ahead of time.
- Define the criteria for success as well as failure.
- Select the suitable and appropriate methodology and be sure that the data is well organized and relevant internal and external factors.
- Validate models with the help of the predefined success and failure criteria.

BA is serious for remaining competitive and achieving the goals. When an individual gets BA best practices in place and gets buy-in from all shareholders, the organization will get an advantage from the decision making which is derived with the help of data analysis.

6.8. WHAT IS INTERNET OF THINGS (IOT)

Internet of Things (full form of IoT) is an ecosystem of associated physical objects that can be accessed with the help of the internet. The "thing" on Internet of Thing could be a person with a heart monitor or an automobile with built-in-sensors, that is, objects that have been given an IP address and have the potential to gather and convey the data with the help of the network, without manual assistance or intervention.

The fixed technology in the objects that assist them in communicating with the internal states or the external environment, which as an outcome, impacts the decision taken.

IoT can associate the devices which are embedded in several numbers of systems to the internet. When the device/object can represent themselves digitally, they can be regulated from any place and at any time. The connectivity then helps to clutch more data from different places, making sure that more ways in order to increase the effectiveness and enhancing the safety and Internet of Thing security.

IoT is a transformational force that provides assistance to companies and organization in order to increase or elevate the performance with the help of the IoT analytics and IoT security in order to deliver the better outcomes.

Businesses in the utilities, insurance, transportation, oil, and gas, manufacturing, infrastructure, and retail sectors can reap the advantages of the IoT with the help of making more informed decisions, which is supported by the torrent of interactional and transactional data at their disposal.

Figure 6.9. Basic significance of internet of things (IoT).

Source: Image by Pixabay.

IoT platforms can help organizations reduce costs through improved process efficiency, asset utilization, and productivity. With improved tracking of devices/objects using sensors and connectivity, they can benefit from real-time insights and analytics, which would help them make smarter decisions.

The growth and convergence of data, processes, and things on the internet would make such connections more relevant and important, creating more opportunities for people, businesses, and industries.

6.9. MERGING DATA ANALYTICS AND IOT WILL POSITIVELY IMPACT BUSINESSES

Data Analytics has a significant role to play in the growth and success of IoT applications and investments. Analytics tools will allow the business units to make effective use of their datasets as explained in the points listed below.

Figure 6.10. Merging data analytics and IoT in order to get positive impact on business.

6.9.1. Volume

There are huge clusters of data sets that IoT applications make use of. The business organizations need to manage these large volumes of data and need to analyze the same for extracting relevant patterns. These datasets along with real-time data can be analyzed easily and efficiently with data analytics software.

6.9.2. Structure

IoT applications involve data sets that may have a varied structure as unstructured, semi-structured, and structured data sets. There may also be a significant difference in data formats and types. Data analytics will allow the business executive to analyze all of these varying sets of data using automated tools and software.

6.9.3. Driving Revenue

The use of data analytics in IoT investments will allow the business units to gain insight into customer preferences and choices. This would lead to the development of services and offers as per the customer demands and expectations. This, in turn, will improve the revenues and profits earned by the organizations.

6.10. APPLICATIONS OF THE INTERNET OF THINGS

There are many applications of IoT. Top 10 Internet of Things (IoT) use cases are discussed below:

6.10.1. Smart Home

Whenever an individual thinks of IoT systems, the most important and effective application that comes in their mind is the smart home, ranking the highest IoT application on all the channels.

The number of individuals who are searching for smart homes increases every single month by about 60,000 individuals. The other interesting thing is that the database of smart homes for IoT analytics comprises two hundred and fifty-six companies and start-ups as well.

It has been observed that in the present times, many companies are actively involved in smart homes, as well as similar applications in the field.

The amount of funding that is estimated for smart home start-ups surpasses $2.5 billion and it is increasing at a quick rate. The list of all the start-ups comprises prominent start-up company names, like AlertMe or Nest, as well as several numbers of multinational corporations, such as Haier, Philips, or Belkin.

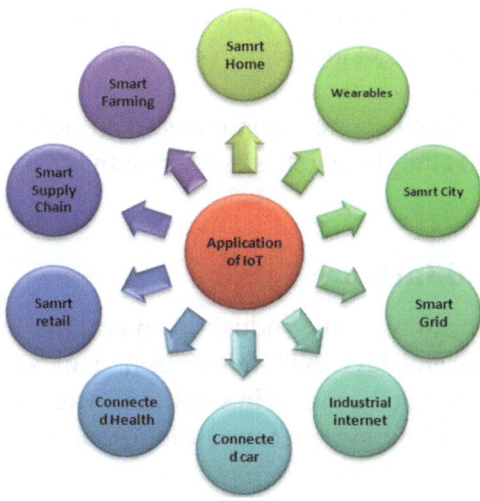

Figure 6.11. Application of internet of things (IoT).

6.10.2. Wearables

Wearables continue to be the hot topic among potential IoT just like that of smart homes. It has been observed that every single year, consumers all over the world wait for the release of the latest Apple smartwatch.

There are a lot of other wearable devices apart from this. These devices help an individual in making their life easy. Some of these devices include, Look See bracelet, the Sony Smart B Trainer or the Myo gesture control.

6.10.3. Smart City

As its name suggests, smart cities are a great invention or an innovation. It spans a wide variety of use cases, starting from the water distribution and traffic management to waste management and environmental monitoring.

It tries to eliminate the discomfort as well as problems of the individuals who live in cities. This is the reason why it is so famous. IoT solutions that are offered in the smart city sector resolve several city-related issues, including of traffic, decreasing noise and air pollution, and helping to make safe cities for people who are living in the cities.

6.10.4. Smart Grids

Smart grids are the other area of IoT technology that stands out. Basically, a smart grid promises to get all the data or the information on the behaviors of consumers and electricity suppliers in an automated fashion in order to enhance the effectiveness, economics, and reliability of electricity distribution. 41,000 Google searches every month is a proof of the popularity of this concept.

6.10.5. Industrial Internet

One simple way to think of the Industrial Internet is by looking at all the connected machines and devices in industries like oil, gas, power generation, and healthcare. It can also be useful in situations where unplanned downtime as well as the failure of the system can result in situations that are threatening to life.

A system that is embedded with the IoT tends to involve all the devices like fitness bands for monitoring the heart or some other smart home appliances. In addition to it, these systems are functional, and they are easy to use. Also, these systems are not reliable because if a downtime was to happen, they do not typically create emergency situations.

6.10.6. Connected Car

Connected car technology is a huge and an extensive network that includes various sensors, embedded software, antennas, and technologies. Connected cars help in communication to navigate in the complex world. This car helps to share internet access, data with the other devices both inside and outside the car. It has the responsibility of making decisions with reliability, correctness, and speed.

Connected cars also have to be reliable. All these requirements will even become even more serious when the individuals give up the control of the steering wheel as well as the brakes to the autonomous automobiles that are being tested on the highways.

6.10.7. Connected Health (Digital Health/Telehealth/Telemedicine)

IoT (Internet of Things) has several numbers of applications in the field of healthcare. These applications of IoT are from remote monitoring equipment to advance equipment and smart sensors to equipment integration. It has the potential to enhance the way how physicians deliver care and it helps in keeping all the patients safe as well as healthy.

Healthcare IoT can also permit the patients to spend more time communicating with their doctors. This can help in boosting the patient engagement and satisfaction. From personal fitness sensors to surgical robots, IoT in the field of healthcare brings innovative tools that are updated with the latest technology in the ecosystem.

These innovative tools help in developing better healthcare for people. In addition to it, IoT helps to transform healthcare and offer pocket-friendly solutions for both the patient as well as the professionals of the healthcare field.

6.10.8. Smart Retail

It has been observed that retailers have begun to adopt IoT solutions. They started using IoT embedded systems across a number of applications that help in enhancing the store operations, increase in the purchases, decreasing the theft, enabling inventory management, and improving the shopping experience of the consumer.

Physical retailers can compete against the challengers who are available on the online platforms more strongly.

This is possible because of IoT. In addition to it, they can also regain their lost market share and appeal to more consumers into the store. Therefore, making it easier for people to buy more products while saving money.

6.10.9. Smart Supply Chain

For a couple of years, supply chains have already been getting smarter. It provides solutions to many of the problems such as tracking of goods while these goods are on the road or in transport or assisting suppliers to exchange inventory information. These are some of the popular offerings of the smart supply chain.

With an IoT enabled system, factory equipment that includes embedded sensors communicate data about various parameters, like temperature, pressure, and utilization of the machine. In addition to it, the IoT system can also process workflow and it can also change the settings of the factory equipment to improve the performance.

6.10.10. Smart Farming

Every so often, smart farming is overlooked in the applications of IoT. Though, because the number of farming operations is generally inaccessible or remote and the large number of livestock that farmers work on, all of this can be monitored or supervised by the IoT. It can also change the way farmers operate from day-to-day. But this concept that is smart farming is yet to reach a large-scale attention.

Nonetheless, it remains one of the applications of IoT that should not be undervalued. Smart farming has the potential to become an essential application field, particularly in the agricultural-product exporting nations.

6.11. CONCLUSION

BA is one of the emerging technologies in the modern world. As there is a plethora of data that exists all over the world and has some insights hidden in it, it has become necessary to pass it through the sieve of requirements.

This process of passing the unstructured data from the sieve of requirements is BA. BA is the process of collecting, structuring, and analyzing huge data sets. After analyzing and obtaining the insights from the huge data sets, these insights can be further used for transforming the businesses. BA is the requirement of every business in the modern world.

There are three different types of BA and these three types can be used according to the requirements of the business organizations. Predictive, Descriptive, and Prescriptive analytics are the three types of BA.

To use these three types of analytics, there are a number of tools that are available for usage in the modern world. However, there are 5 major types of tools that are usually used. These tools are Sisense, Qlikview, Clear Analytics, MicroStrategy, and PentahoBI.

The insights obtained from BA then can be further used to align physical objects and along with the technology. Aligning the physical objects with the internet is called the IoT (IoT). IoT is the future of the world.

Everything that has been automated with the help of the internet involves usage of IoT in the first place. Moreover, as the human race has involved internet in every aspect of life, IoT is the linking bridge between the internet world and human life.

Combining IoT and BA can prove to be very helpful as it explains how the various insights from the data present in the world can be helpful in linking the machines through the internet according to the needs of human beings.

REFERENCES

1.   Abhinav, R., (2018). Business Analytics: Tools, Applications & Benefits. [online] upGrad blog. Available at: https://www.upgrad.com/ blog/business-analytics-tools-applications-benefits/ (accessed on 10 March 2020).

2.   Clark, J., (2016). What is the Internet of Things, and How Does it Work? [online] Internet of Things blog. Available at: https://www.ibm. com/blogs/internet-of-things/what-is-the-iot/ (accessed on 10 March 2020).

3.   Happiest Minds, (2020). Internet of Things (IoT) – Happiest Minds Insights. [online] Happiest Minds. Available at: https://www. happiestminds.com/Insights/internet-of-things/ (accessed on 10 March 2020).

4.   Micro Strategy, (2020). Business Analytics: Everything You Need to Know. [online] MicroStrategy. Available at: https://www.microstrategy. com/us/resources/introductory-guides/business-analytics-everything-you-need-to-know (accessed on 10 March 2020).

5.   Molly, G., (2016). NGDATA | What is Business Analytics? See Benefits and Applications. [online] NGDATA. Available at: https://www.ngdata. com/what-is-business-analytics/ (accessed on 10 March 2020).

6.   Quant zig, (2019). Importance of Business Analytics, Advantages of Business Analytics | What is Business Analytics. [online] Quantzig. Available at: https://www.quantzig.com/blog/importance-business-analytics (accessed on 10 March 2020).

7.   Raut, S., (2011). So What is Business Analytics and Its Various Components? [online] SmartData Collective. Available at: https:// www.smartdatacollective.com/so-what-business-analytics-its-various-components/ (accessed on 10 March 2020).

8.   Rinu, G., (2018). Top 10 Applications of IoT—DZone IoT. [online] dzone.com. Available at: https://dzone.com/articles/top-10-uses-of-the-internet-of-things (accessed on 10 March 2020).

9.   Sachdeva, R., (2014). 6 Must-Have Features for a Business Analytics Tool. [online] Grazitti Interactive. Available at: https://www.grazitti. com/blog/6-must-have-features-for-a-business-analytics-tool/ (accessed on 10 March 2020).

10. Schniederjans, M., Schniederjans, D., & Starkey, C., (2014). What Are Business Analytics? | 1.1. Terminology | InformIT. [online]

Informit.com. Available at: http://www.informit.com/articles/article. aspx?p=2206307 (accessed on 10 March 2020).

11. Tony, J., (2018). Role of Data Analytics in Internet of Things (IoT) | Fingent Blog. [online] Fingent Blog | IT Solutions Blog | Ideas to Motivate Business Growth. Available at: https://www.fingent.com/ blog/role-of-data-analytics-in-internet-of-things-iot (accessed on 10 March 2020).

New Age Customer Acquisition with Retail Analytics

CONTENTS

This chapter is about how retail analytics has changed new age customer acquisition. As retail marketing is rising and is so challenging in terms of attracting the customers. Nowadays, customers have lots of retail options to find their required product.

For retailers, it is a bit difficult to acquire new customers as customers are having lots of options for their products. Retailers need to understand that there is a need to give importance to each and every point related to customers and the product. There are different prospective that one must keep in mind at the time of customer acquisition.

7.1. INTRODUCTION

With an increase in technologies, retail sector has changed a lot. As retail businesses increasing, different technologies are helping retail sectors to grow. The needs of a customer are increasing with the availability of products. Retailers are trying their best to gather a greater number of customers to get a large profit.

Traditionally, customers were more focused towards on store shopping. That time customers acquisition was not that easy. Customer's acquisition was considered a little difficult, as the different retailers are providing different offers, a product with quality differences. It was not considered easy for the customers to get their desired products at the cost they want and with good quality.

Figure 7.1. Customer acquisition.

Source: Image by pixabay.

Whereas nowadays, customer acquisition is also not easy. As customers are more aware about the type of products they need. As with the increase in the retail options, the new emerging brands have made these customers more brand conscious. So, to make the customers understand and believe in their product is not easy.

Most of the retailers with a big brands are using different technologies, like Big Data, to acquire new customers. As Big Data helps in providing information regarding the customer choice, feedbacks for old items, information regarding the new product they are searching for, and much more information that help the retail to understand the requirement of the customers and to maintain their old customers.

There are different researches that is being done to explain the importance of customer acquisition and the ways to do that. There is a certain point that every retailer must keep in mind. These points are very necessary to gain new customers and retain old customers. So, new customer acquisition is important to maintain customer retention.

7.2. DEFINING RETAIL ANALYTICS

Retail Analytics is the process of providing analytical data related to sales, inventory, customers, and other important aspects crucial for merchants' decision-making process. The retail analytics data is used for market decision making. This discipline covers several fields and creates a broader picture of retail business' health and sales along with the areas of improvement and reinforcement.

Essentially, retail analytics is used in making better choices, run businesses more efficiently, and deliver improved customer service analytics. The retail analytics approach is beyond the superficial data analysis (DA) that uses data mining and data discovery to sanitize data sets in order to produce actionable BI insights that can be applied in the short term.

Also, in order to create better snapshots of the target demographics, companies are using retail analytics.

Retailers nowadays can identify its potential customers by harnessing sales data analytics on the basis of different categories such as age, preference, buying patterns, location, and more.

The field of retail analytics has not only emphasized on just parsing data but also gave emphasis on what is the information needs of the businesses, how the data can be best gathered and how the data can be best used in an efficient manner.

Since companies are prioritizing retail analytics that focuses more on the process and not much on data itself, companies are getting stronger insights and stand in a more advantageous position to succeed while predicting the business and consumer needs.

Figure 7.2. Retail analytics.

Source: Image by piqsels.

7.3. DEFINING CUSTOMER ACQUISITION

Figure 7.3. Customer acquisition.

Source: Image by Pixabay.

Customer Acquisition is defined as the process of gaining new customers and retaining the older ones. It also involves the conversion of existing prospects into new customers. To acquire new customers, it is important to persuade consumers to purchase company products and services.

Customer Acquisition plays an important role in businesses, as it is one of the major factors that contribute towards achieving the organizational objective. Customer acquisition helps businesses to evaluate the value customers bring to the organization or businesses.

Different sets of methodologies and systems are used to maintain or manage customer acquisition. Various marketing techniques are used to manage customer's inquiries and prospective. Customer referrals, loyalty programmes are some of the marketing strategies that are used to obtain successful customer acquisition.

It is important to have a link between advertisement and customer relationship management (CRM). This is considered as a critical connection that helps businesses to acquire and have positive effects on the targeted customers.Customer acquisition varies according to specific business situations. The business focuses on acquiring customers at less cost, acquiring as many customers, acquiring customers who are business-oriented and indigenous, acquiring customers who focus on new technologies or new business channels.

Customer acquisition process considers different aspects:

First, it is important to concentrate and focus on customer psychology. Customer feelings, thoughts, thinking, and requirement are some of the basic things that a business focus on. This type of strategy helps the representator to select and represent the right product in front of the customer.

Concentrating on how the customer is influenced by the surrounding. Business culture, technologies, and media are some of the aspects that a representator should focus on while acquiring new customers.

Analyzing customer behavior and tendency during the purchase of a specific range product.

Understanding and studying the customer knowledge limitation about the product. The understanding of the product helps customers to buy things and make decisions accordingly.

Best strategies should be used to convince customers. Using new strategies will involve new customers and will improve marketing campaigns.

Customer acquisition techniques changes with the change of technologies. It is important to upgrade and optimize the traditional ways of marketing channels that effects consumer acquisition. The involvement of new technologies and strategies is very important in acquiring new customers. It is also important for businesses because that helps them to remain in the competition.

Acquiring customers depends on organizational behavior. It depends on how the organization is building a relationship with the customers. If the relationship between supplier and customers is healthy and effective, then the company is considered to have high revenues. A healthy relationship brings confidence in customers to buy a greater number of products from the specific supplier.

It is also possible that a satisfied customers can seek to buy specific products different from the regular product from the same supplier. For example, a customer has a car insurance with a specific service provider and that company is providing many other benefits with the regular insurance policy.

This marketing strategy makes the customer get its other vehicle insured from the same company. There is also a possibility that the same customer gets the same company insurance again and again. These strategies make company benefits and result in good business growth.

It is important to understand the nature of the response to acquisition. It is the key aspect that creates an impressive opinion in customer minds. To serve customer with the best information, it is important for the suppliers to have a prompting nature, should be responsive and experienced.

The acquisition of customers is not an easy task for the suppliers. But if the strategies, understanding, and response of the supplier are good, then the customer tends to bend more towards their products. Taking an example of a company selling the latest technology speakers.

If the representator is unable to describe about the product and is also unable to answer the questions of the customer. Then this will lead the customer to go out and look for other options. This way the company loses its customers and the relationship between them breaks.

For the organization, it is very important to acquire new customers. It is considered as the biggest task for them. So, to acquire a newer customers, it is important to identify critical approaches. This includes a greater number of customers to get acquire at a very low cost.

Promotional campaigns are being organized, which is considered as the best strategy to helps the businesses acquire new customers. Campaigns are meant to be efficient and targeted to the customer.

Customer encouragement is one the best way to have new customers connected with the supplier. Customers who are already supporting the company, help to promote the product of the company and that is free of the cost. Satisfied customers help the company by bringing new referrals by simply sharing their experience towards that particular supplier.

Customers are sharing their positive responses towards the supplier, helps companies to acquire new customers easily and with this they do not fear to lose customers who are already part of the organization.In terms of revenue, the organization should have a balanced number of customers acquired and the number of customers who got disengaged and get diverted to other company products. Which directly effects the growth of the organization.

7.3.1. Customer Retention

With customer acquisition, it is important to have customer retention. Customer retention is defined as how companies and organization are maintaining their old customers with maintaining good relationships with the new customers. Some researchers have outlined some of the philosophies oriented to retention:

- Company must allocate some marketing activities that are required to generate high profit in the company.

- Sellers take advantage of those customers who feel good and smart about their choices. Sellers provide them promotions, that engage the consumers and make them feel good about it.

- Keeping customers active is an important thing, if a company does not pay attention to their customers then there will a loss of customers.

- Marketing with market data is considered as highly evolved and valuable. But it must be backed and forth between the consumer and the seller because it is important to listen what customer wants to say about their products and the services to have a better understanding about the market.

7.4. COMPONENTS OF RETAIL ANALYTICS

According to The Retailers Association of India (RAI), a successful retail strategy covers the following six areas:

7.4.1. Predictive Modeling

Predictive modeling or predictive analytics is the process of using data analytics to make predictions based on the data. The process uses data along with analysis, statistics, and machine learning (ML) techniques. This form of the model creates a predictive model for forecasting future events.

In predictive modeling, historical data is used to predict future events. Basically, a mathematical model is made using historical data that helps in capturing important trends. Now, that predictive model is then used on current data to predict what will happen in the future and helps in taking action for optimal outcomes.Predictive modeling is very popular in recent years due to advances in supporting technology, particularly in the areas of big data and ML.

7.4.2. Big Data and Hybrid Architectures

The data world is heterogeneous. There is a huge competition on analytics and the adoption of a whole host of new data technologies and accompanying best practices is what needed. The convergence of structured data and unstructured data through the integration of data in different types of apps, social media and other channels.

7.4.3. Cloud Analytics

Cloud Analytics allows its users to access the data faster. Cloud analytics is highly scalable and has the capacity to store and access relevant information. It helps the businesses who are using big data in their day to day work in achieving their goals.

Cloud analytics are having some elements of analytics that are data sources, data model, processing applications, computing power, analytic model, sharing, and storage.

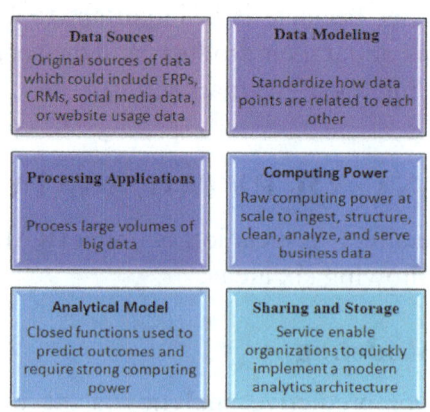

Figure 7.4. Elements of data analytics.

There is a different type of Cloud analytics that are used nowadays:

7.4.3.1. Public Cloud

In public cloud analytics, Virtual machines, storage, and data processing are some of the organizational applications used as a service. Public cloud is available for common people, where the system is being shared but not data. This process helps companies to reduce the cost and efficient in maintaining the IT managements.

7.4.3.2. Private Cloud

A private cloud is dedicated to a single organization. This type of cloud is preferred by a single organizations as their private property. They are accessible only by company workers. They serve as an extension of existing organization IT infrastructure.

Private cloud is being accessed only when there is an issue related to data piracy and security. Companies who believe that their data should be preserved and should be accessible by limited people, then private cloud is used. This type of cloud analytics is a bit costly then any other cloud analytics.

7.4.3.3. Hybrid Cloud

Hybrid cloud is a combination of both the clouds that is, the public and private both. In the hybrid cloud, there are some data that is not important to keep private and can be available for all the people. But there can be some data that need to be share between specific people. This type of cloud analytics helps to maintain both the data, either public or private, and benefits in IT infrastructure demands.

7.4.3.4. Advanced Visualizations

Advance virtualization is the act of creating a version of something. It is defined as the art and science that is used for making the function of an object or resource simulated or emulated in a software that is identical to the corresponding realized objects.

In other words, using an abstraction that will make software to look and behave like hardware. Advanced visualization is beneficial as they are flexible, reliable, and in applications. This advance virtualization increases productivity and side by side save results from the reduction of hardware and space requirements.

7.4.3.5. Real-Time in Memory

It is a platform where users run their queries and interact with the data that is stored in the main memory instead of a hard disk. This type of approach works faster to the queries response, they act high in performance, reduce operational cost, encourages self-service practices. It makes faster extraction of data with real-time DA. There are several objectives used in the new era.

7.4.3.6. Self-Service Analytics

Self-service analytics is when a customer makes decisions based on their own queries. Self-service helps the user to decide and choose the process and products based on their own analysis. In this type of analysis, the user can easily analyze, extract, and build their own reports and help the user to maintain their old reports. This doesn't require training.

This is user friendly, as it allows to build reports on the basis of exact information. This type of analysis reaches beyond the pre-defined data.

7.5. USES OF RETAIL ANALYTICS

There are several retail analytics examples that are relevant to a variety of companies. One of the important benefits the retail analytics delivers to companies is that it helps in optimizing the inventory and procurement. Nowadays, businesses use historical data and trend analysis to determine the products companies should order and determine the quantities instead of relying exclusively on past orders.Additionally, companies can optimize their inventory management to predict the products customers need, reduce wasted unused space and associated overhead costs.

Apart from inventory activities, retailers use analytics to identify customer trends and changing preferences by combining the data sets from different areas. Businesses merge sales data with a variety of factors that help them in the identification of emerging trends and anticipate them better. This is closely related to marketing functions, which also benefit from analytics.

Companies are continuously trying to build an improved understanding of individual preferences and having more precise insights by harnessing retail analytics to improve their marketing campaigns. They combine demographic data with information such as shopping habits, preferences, and purchase history of the customers and create strategies that focus on individuals and exhibit higher success rates.

7.5.1. Components of Retail Analytics

According to a new report titled, "Driving Retail Growth by Leveraging Analytics" by consulting firm PricewaterhouseCoopers (PwC) and the Retailers Association of India (RAI), a successful retail analytics strategy, will cover the following six areas:

- Predictive modeling: Developing an analytical model to predict future outcomes and empower business users to make decisions quickly.

- Big data and hybrid architectures: Convergence of structured and unstructured data through data integration across apps, sensors, social media, and other channels.

- Cloud analytics: Highly scalable and easy way to store and access relevant information, which allows users to access more data faster.

- Advanced visualizations: Present data in visually compelling ways, enabling companies to expand BI capabilities extended to their executives and other employees.

- Self-service analytics: Making analytics a more democratic process by allowing users to make decisions based on their own queries without requiring any sophistication.

- Real-time in-memory: A move ahead of the traditional relational database that can help retail analysts to generate deeper insights across the entire value chain of retail operations, including procurement, supply chain, sales, and marketing, store operations, and customer management.

7.6. CUSTOMER ACQUISITION PROCESS

Customer acquisition is based on planning and strategies. Every company makes different marketing strategies to acquire new customers. Customer acquisition strategies are also sometime used as part of the customer acquisition process. Few basic points are used for different types of customer acquisition, but there are also some more effective methods that can be used for some specific types of customers.

Basic customer acquisition plan includes identifying quality potential customers. One of the strategies to acquire customers is to connect the potential customers through call centers and mailing lists. These strategies are used to determine the customer's interest towards a particular product. It also helps companies to determine individual or customer is using which company product.

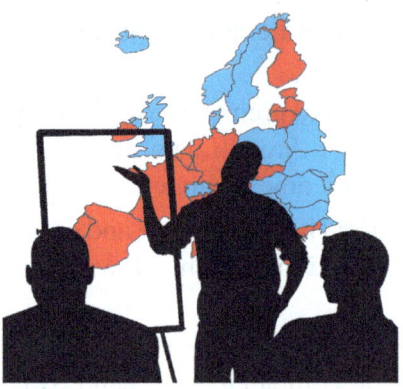

Figure 7.5. Customer acquisition strategies.

Source: Image by needpix.com.

Customer acquisition programs include relationship maintenance and establishment on the basis of the requirement of the customers. To understand customer needs is especially important while offering products and services. As their needs decide the product to be offered by the company or representator. Sometimes the requirement is being identified from the ongoing conversation and interactions with a salesperson.

So, to have a successful acquisition of customers, it is important to have a strategy. Planning is the best way to execute each step. So, before implementing anything, it is important to have strategy and planning.

There are various points that are helpful in creating successful customer acquisition:

7.6.1. Target Prospective

Customer acquisition begins with identifying the potentials and the customers who are important for business success. To acquire new customers, it is important to understand that who are the target customers, who can invest in products, who can understand the product and identifying the customers who can help in business success.

To have a successful acquisition of customers, companies need to find who are the customers, how to reach them, what are the points to discuss, and how to sell the product. This process can help to understand the requirement of the customer.

To find the right customer is very important for businesses. If the customer is right for the company, then that will help the supplier to understand the customer's interest in the product. This also helps to understand which individual is interested in buying a product from the company or using the same product from different companies or competitors.

7.6.2. Reaching Targets

Once the target is being decided, it is important to see the ways to connect them. Several researches are being carried out to connect the target customers. Companies are trying their best to reach their target customers.

Several strategies such as Research, calls, emails, analytical tools are used to connect their targets. If the company is unable to find its targets, and acquisition of these targets is successful, then there is a chance of turning new customers to potential buyers.

7.6.3. Good Supportive Team

It is very important to hire the right staff for the company. The acquisition of customers involves lots of dedication and hard work. So, to have successful acquisition, hiring the right person who can develop a great understanding of the right process and should know how to implement it at the time of acquisition. The employee should be dedicated to their work and responsibility.

Taking an example of a salesman. A salesman should always be active and ready to identify customer requirements. One has to identify the requirement of the customer from the ongoing conversation. They should be able to understand the correct problem and understanding of the customer for a particular product. To have a flawless experience, support teams are required to respond immediately to customer queries.

7.6.4. Customer Acquisition Cost

Determining the cost of the company is going bear during the implementation of the process is important. Companies are believed to spend more than their customers. Acquisition of customers require more cost then the amount of customer gets acquired. This cost is being calculated by the cost that is spend in the acquisition process, which includes sales and marketing cost, divided by the total number of new customers being acquired in a given time frame.

Acquisition cost is used to define customer value. Companies who spend their money to acquire new customers and maintain the old ones, expect a high return on their investments. Their main goal by spending on the acquisition process is to gain a high return on investments (ROIs) value.

Therefore, company investment is directly related to the customer acquisition rate. Every company before investing on acquisition process make sure that their target to achieve great amount of acquire customer is reached.

7.6.5. Creating Product Demand

In the case of big brands, they have a large number of customers who are aware of their products and believe in their brands. So, to acquire new customers for their brand is not a difficult task. But if the company is new or start-up, then those companies suffer lots of problems in the acquisition of customers.

If the brand is small or the company name is not much famous, then it becomes difficult for the salesman and the company to create their space in the market.

It becomes more challenging in today's world, as new technologies are being introduced by big brands. There is a number of companies producing the same product. In the market one can find a product that is available in different brands. Smaller companies are tending to fight for their existence. As to convince customers to buy their product has become very difficult and time is taken to process for small-scale industries.

7.6.6. Build Optimizable Web Pages

The traditional way of customer acquisition was to reach them directly door to door, to make them buy their products. But today's technology has made the customer acquisition process easier. Organizations use Web pages for direct customer acquisition. They describe their products on that web page, including images and the cost with relevant discounts.

This technology has made easy for the customer to connect with the product seller and know about the recent product introduced. It is necessary for the businesses to make a web page in such a way that it is responsive on big screens as well as mobile small screen and it should be user friendly. A web page should be designed in such a way that it should be according to the customer expectation.

7.6.7. New Customer

Traditional technologies are considered not to put a focus on customizing products for the customers. Before, the product was used to be the same for all the customers. But the recent technologies have made it so easy. Today, organizations are selling customize product to their customers. They are selling products according to the need of their customers. Characteristic of a new age customer is defined by:

• Immediate access to a range of choices and options;

• A journey towards a purchase decision is more complex;

• Loyalty is momentary;

• Influence;

• Use the path of least resistance to close a transaction; and

• To sell tricks are required.

Today, companies are using different strategies, that are helping customers to choose a product according to their requirements. Companies are using a strategy of turning things or products according to the requirement of the customers. They are trying their best to acquire customers, by providing them best facilities. This means that companies look beyond traditional approaches to reach out customers requirements.

Figure 7.6. Reaching new targets.

Source: Image by Public Domain Picture.

7.6.8. Content Writing

Content Marketing is one of the ways used to acquire customers at a large scale. It makes easy for the businesses to publish their product details at a large-scale using content writing mechanism. Through content writing businesses describe their products using blogs, social media posts, advertisements, etc. This helps the business to reach a large number of customers easily and make them aware of their product that is being recently launched.

Customer engagement is very necessary, there is some point important for customer engagement:

- Introducing models of engagements that are evolved from customers;
- To deliver consistency in customer experience across the touch point by moving away from the product centricity;
- Simplification of IT infrastructure that will target on reduction of transaction complexity;
- Availability of choices;
- Customer centric roles;
- Convenience matters more than channels; and
- Finding customer interest and motivations.

7.7. RETAIL STRATEGY AS DIGITAL DISRUPTION

Collecting, processing, and sharing of a large quantity of data are responsible for bringing disruption in fundamentals of design of business models. There are several factors impacting retail strategies are:

- Convergence;
- Customer centricity;
- Co-opetition;
- Co-ownership;
- Co-creation; and
- Continues learning.

7.7.1. Convergence

Convergence is the ability to establish a relationship between the customer and the producer. Convergence help in enabling the user to purchase their products and services by creating an omni-channel experience.

This channel provides multi-channels approaches that provide customers to have an integrated shopping experience. Innovations in the sectors such as payment systems, e-commerce, etc., are disrupting the fundamental of the business.

7.7.2. Customer Centricity

To align the structure of customer need, functional boundaries that are in the organization is being broken down. Big data is making easies to analyze customer behavior in detail.

It also customizes the delivery format according to the requirement of an individual consumer through personalizing the format in the way they want. The process of customer-centricity accelerates as big data through a digital medium.

7.7.3. Co-opetition

Co-opetition is a revolutionary thinking that combines competition with collaborations. In earlier times, businesspeople were fighting for limited resources. They were more focused on providing resources to their customers.

But in today's era, suppliers, and retails do not focus on limited resources, but they put their main focus on working together to get profit from the customer value that is being delivered. Some of the examples that explain the best collaboration between retails are Flipkart and Future Groups.

7.7.4. Co-ownership

Co-ownership in retail is the biggest approach that is used to monitor and control the quality of a product. Co-ownership helps by providing a franchise agreement to the person who is eager to take the responsibility of maintaining the product quality. This allows the organization to know the exact requirement of the shopper.

This does not lead the shopkeeper and the buyer to compromise on the brand of the product. It also helps in sharing the burden of their partners. Co-ownership is a very common aspect that is used nowadays to provide branded products across the world, even in smaller cities. Co-ownership provides the best quality of products to the customers.

7.7.5. Co-Creation

Co-creation is creating a group of individuals that work as an advisor. Different individuals like consumers, vendors, designers, suppliers, retailers, and subject experts. Co-creation is when these groups of people advise or share their experiences related to the product on social peers and let other customers get benefits from their feedbacks.This feedbacks are considered very important in today's era. Every customer nowadays believe that before buying a product, the reviews of a customer who already bought the product is very important. This feedbacks help in refining the model delivery on a real time basis.

7.7.6. Continuous Learning

Using Artificial Intelligence (AI) and Cognitive Modeling for self-learning has made customer shopping easy by providing them feedbacks of the customers from the older customers who already bought them. This brings satisfaction and loyalty to the customer in real time.

Traditionally, customers were using different ways that are used at the time of buying products are superior customer service, high operational efficiency. Digital analysis has made customers to choose best for them relating the past customer experiences. This also helps retails to enhance delivery and helps to improve their quality according to the customer need.

7.8. HOW RETAIL INDUSTRIES TRANSFORMED BY NEW AGE CUSTOMER ACQUISITION

The process of customer acquisition has changed a lot. The way of dealing with customers has changed with time. There are many factors that are responsible for the rapid growth of the industries. These developments have raised per capital income, evolved consumerist sensibilities of urban and semi-urban areas, changed the lifestyle.There are several opportunities that are used by the retail landscape:

7.8.1. More Efficient Mapping of Consumer Demand

With the increase in brand quantity, people are becoming more conscious about the brands. In many countries, emerging brands have affected the people or consumers to focus more on brands. This trend emerged from the bigger countries to smaller countries. This rising demand for branded products has made the local market in a loss.

Demanding branded products, increased the incomes of retailers and increased the growth of consumers' purchasing power. Brand emergence has changed the mindset of consumers. It made the customers more brand conscious. Customers are forgetting about their native products and focusing more on branded products. This made retails of local business to be in loss.

This brand consciousness is not limited within metropolitan cities, but also in semi-urban markets, customers are asking for the specific brand products. Customers are now more up to date with the latest products and brands.

New age customers or consumers are being developed by the rise of e-commerce across the country. Products such as goods and services are being sold on online retails applications. The e-commerce has made options for consumers, to select their brand products according to their requirements.

The development of e-commerce has also increased competition in the marketplace. Retail brands were looking to thrive in a way that rapid transformations digital landscape helps in reading the requirement of the consumer. To do these transformations, retailers require real time access to enterprise data that may be available in different places.

In a market like America, India which are diverse, large, and fragmented areas, transformation becomes a challenge for most of the retails. Because in these areas the relevant information is bounded by an individual across the various environments. To reach people to gather products is considered difficult because there can be online and offline availabilities.

7.8.2. Market Penetration Beyond Tier-1 Geographies

Organizing retail in tier-1 and tier-2 markets is considered as difficult. It is considered that in semi-urban and rural areas these facilities are absent. Though most of the county population lives in these geographical areas.

An organization who wants to expand their businesses in these tiers, Investors in the organized retail space need to invest heavily. To setup their businesses they require to make the heavy investment of capital, time, efforts, and resources.

In these areas, it is difficult for businesses to have data regarding the need of the customers. Organizations do not have enough data about the target markets and the customers, which makes it difficult for them to convert or translate accrued data into actionable data.

These actions lead businesses to miss the strategies that are required to set up businesses in these areas. This impacts the profitability of the business. In this case, many retail brands take the risk. This risk can be profitable, and some time can be non-profitable.

7.8.3. Disorganized Supply Chain Management

Supply change management is a system used by small and large organizations to get products to consumers. Supply management is done to obtain raw materials, manufacturing, and delivering the final product to the customer. A well-organized supply chain management system involves optimizing operations functionality to be fast and efficient.

Data-driven logistics is another factor hindering retail business owners. This inhibits the owners from gaining profits. This happens when there is an absence of robust. It is considered that most of the data that is being generated at each step of the supply chain, is not in an organized way nor in a digitized way.

To maintain Streamlined and agile distribution channels, it is important to digital optimization. The lack of optimization is considered as challenging in establishing these criteria across the county.

Maintaining the supply chain is important for improving customer services. Which includes, customer expectation of receiving correct product and quality on time. Secondly, the product is delivered on the right hands and in the right location. This also helps in maintaining data related to customer reviews. This needs to be done as soon as possible.

7.8.4. Enter Data Analytics

In most businesses, data analytics is used to overcome the gap across several verticals such as sales, inventory, and services. Retail Brands can leverage data driven, to understand potential consumers based on various geographies. This helps them to expand their strategies according to the data captured.

Geospatial visualization and analytics provide a unique ability to the businesses to decide the correct place for stores. It also stores the performance of customers and makes consumer analysis.

7.8.5. Leveraging New-age Analytics to Streamline Retail Data Operations

Traditionally, the Analytics process requires specialized data, that is being gathered by the professionals. They analyze and create relevant data from the raw data. This process is considered as time consuming and lots of labor is required to create in house data team. This leads to data funnel in centralizing data operations.

Data integration to the analysis of data and bringing insights to the business user via chat services, provides retail sellers to the solution for end to end management. This process enables retail businesses to extract information across different places, formats, and storage.

In every context, it is important to provide a solution to end users with relevant information that is important for businesses during decision making. This brings down the time taken from insight to action. But it also allows the end-user to freely explore various data combinations in any direction to unlock hidden business value.

Figure 7.7. Customer acquisition insight.

Source: Image by Flickr.

To understand the retail industry, it is important to understand data analytics. As data analytics needs should be made as a part of everyday workflow. It is important for the businesses to become comfortable with the data, whether it is in the form of reading, working or analyzing and arguing with data provided. By doing this, ensures that the data-driven business framework will yield robust dividends.

7.9. RETAIL HABITS ADOPTED BY THE NEW AGE CONSUMERS

Figure 7.8. Customer satisfaction.

Source: Image by the Blue Diamond Gallery.

With the transitional shifts in different countries, customers have changed or reshaped their retail habits and preferences. Shopping has shifted from malls or shops to online using mobile phones. In the last few years, the change in the customer ways to buy things is being affected by digital disruption. The purchase pattern of the consumer has changed.

Millions of consumers have acquired new age consumer habits. Customers need to gain satisfaction with the product offered by the retails. This made to collapse most of the routine purchase patterns. Some studies are being carried out to explain global consumer insights. These studies are focusing on the changes in customer behavior related to new market technologies. As there have been many changes in new age consumers, their habits and the way to attract these customers.

It is noted that customers are now more brand conscious, so to reach the customers, one must be aware of brands. These researches have also pointed out that the old technology of buying products by visiting stores is still considered on a large scale. Customers still consider buying a product from stores directly. They also said that brands have to pay more attention to smartphone and their investments to these AI are also very high then stores.

Below are some of the retail habits that are being adopted by the customers:

7.9.1. Retail Store Are Still Preferred

Retailers have a fair knowledge of today's era. They have understood that this era belongs to the Omnichannel. So, they have a diverse marketing strategy. The statistics of the survey are considered as interested. The destination for shopping matters the most.

Today's people are still connected with the traditional way of shopping. They still believe in brick and mortar stores. This shows that the majority of people are still connected with this prospective. People are heading toward the online ordering services and pick up store services, because the traditional way of store shopping includes standing in a long ques.

Some researchers said that million would thrive on convenience and time factors and visiting the store will always be considered unless the shopping is unplanned. Some factors such as touch and feel, where customers believe in buying things based on the quality of material after touching it on stores. Because of this reason customers prefer the Omnichannel presence and ensures the brand quality by looking for availability on both offline and online.

7.9.2. Rise in Mobile Payments

Today, as there is an increase in the use of online store shopping, there is also an increase in payment methods. Everyone in the world is moving towards cashless services. Digital transactions have an overwhelming impact on the use of mobile payments.

In some surveys, customers said that they believe in paying through mobile phones payments methods to complete their transactions. They said they prefer mobile transaction for online payments as well as for paying in stores.

Some researchers said that especially e-wallets have become extremely important to capture customer preference. Through retailers understand the customers spending behavior. When a value is added to e-wallet, it means sales. This is considered a positive trend which will make more easy business.

Figure 7.9. Payment method preference.

Source: Image by Payrollservices.com.au.

7.9.3. Delivery Speed Plays a Key Role

Instant gratification is mostly used process by the customers. This means that customers do not think about the ways of buying. Simply, this means that as customers require something or want to buy something can order directly. To buy something customer doesn't need to think about it until their next planned outing.

It is even easy to buy things or order something, as one must ask their digital assistant to make their purchase. Example Amazon Echo or Google Home. Many customers are even paying extra charges for the delivery of their products on the same day.

7.9.4. The New Influencers

Recent studies said that, the new age customers are different from old generation customers. It is considered an important factor in understanding the peers of customers. Recent studies have explained that social media plays a key role in helping a brand to connect with the millions of customers. It is more considered as the brand emails or emails from retails do not connect well with customers. They tend to reject the intrusive sales pitch.

People at high positions have said that from the time social media came into the influence or existence, it is considered as an influencer, as people on social media are looking constantly for reference points. The social media platform is considered as a huge influencer. The social media platform is more often used for guidance.

Now, it is considered that retailers should focus on entering the social media platform for marketing without involving much interference. It is suggested that retails should find a marketing model, which will help to get the attention of the consumers. With this, they should have a relevant place with a purposeful campaign. When once this is achieved, it will have a brand and consumer connect.

Industries experts have read in some surveys and accepted that companies should respond to new customer's habits. This is possible only when they start focusing and investing accordingly. New business practices are involved in supporting the investments. This will help companies with the flow of new customer behavior.

7.10. IMPORTANCE OF BIG DATA FOR RETAILERS

Figure 7.10. Big data for customer acquisition.

Source: Image by Flikr.

There are some ways through which Big Data is helpful in retail business:

7.10.1. Identifying and Creating Client Profiles

Recent technologies have made easier for the retailers to have a better way to identify the customers and also to choose the right products for them. Customer transaction history, basket analysis, loyalty programs and social media interactions have made customers segmentation much more refined and data have driven.

Personality points, demarcating faceless mass into slots are some of the points that make a study of buyer data according to the purchases using Big Data managements. Transactions reports and loyalty plans are combined to carry out relevant data and bring that into action.

This makes easier for retailers to get the whole view of the customer. This makes them customize the product according to their past preference or similar to the product which is bought before.

7.10.2. Price Optimization

Online retailing is based on the pricing system. The price of the product depends on many factors like marketing demand, inventories, competitor's pricing, seasonal products. Data analytics plays a vital role in determining the price of the product. The price of the product depends on the current scenario like product demand, product price on other stores.

These algorithms track the demand, inventory levels, and competitor's activity. These automatically respond to the market that changes in real time and allows actions that to be taken that depend on insights in a time saver manner.

Mark down optimization is defined as determining the price of a particular product, which is dropped due to some reasons. Algorithms help in determining the prices of a product that needs to be dropped. According to the analytics, prices are being dropped down as most of the retailers would reduce the price at the end of a buying season to achieve a particular product line, at the time when the demand is almost gone.

But ML has made easier for retailers to do the pricing details. This tells the retailers to adjust the price of a product in real time. Recommendation and offers are being sent to specific customers who might have already shown interest in that product or might have bought the product earlier.

7.10.3. Generation of Customer Loyalty

Customers' expectations have reached so high that they feel that they should be treated royally. They are expecting a lot from retailers. They expect retailers to send information regarding every product in the way they want. They want retailers to understand the requirement of customers, recommend products accordingly, and provides services that suit them well.

They want to be informed at every stage, from ordering their products, shipping details and product delivery feedbacks. This is not an easy task for the retailers to keep in mind, the detail of varied customers they serve. Big data analytics helps retailers in recommending the right product to a customer. With the help of Big Data retailers come up with targeted marketing campaigns to reach specific segments.

Big data helps them to understand the customers in a better way, as it provides understanding a path to purchase or to understand buying patterns. This helps them in reaching out to their customers at each and every step to close the sales cycle.

7.10.4. Prediction of Demand

Recent advance tools availabilities have made easier for the retails to get up to date with the trends of the industries. Forecasting demand has become a very efficient tool nowadays. Though this retailer can easily find whether the product is in demand or not. It also helps in predicting the demand during a certain times of the year, in a particular city.

It also helps in adjusting the inventories. Retailers are focusing on gathering information from social media platforms to understand customer demand and preferences. They also find whether the product is getting a negative response or positive response or neutral responses in the market.

Big data helps in predicting the current requirement of the customer and to understand the need of the customer with the help of data that defines the previous action taken by the customers.

7.10.5. Management of Inventories

Trend forecasting algorithms helped in sorting the buying data that will help in analyzing the need of the market. This helps the marketing departments to promote the need of the customers and help the retailers to decide what to share and what not to share. When one it is clear about the requirement of the customers, trend they are following, then the retails starts focusing on the sectors where there will be high demand.

This helps in gathering demographic, seasonal, occasions data, economic indicators that help in building a picture of buyer behavior across the target market. This helps in managing inventories in a better way.

7.10.6. Identifying Opportunities for Highest ROI

Customer interactions can have a larger impact on the existing relationships and potential relationships. Applying new ideas to the whole sales forces can lead to a risky endeavor. As one wrong decision can lead to immediate loss of profit at an instant that can last for a longer time. Big companies or leading companies have found the best way to isolate the cause and effect relationship.

They use to test and learn approaches for strategic shift and key performance indicators (KPIs). This means trying an idea with some sets of customers and comparing the performance of the test group to the well-matched groups' performance. After gathering the best understanding of the current and potential customers, retails calculate and predict the risk filters. They also assume their responses in marketing campaigns.

7.10.7. Detecting Frauds

Big Data is effectively used to detect frauds. Frauds are being analyzed by data captured from daily transactions and activities. Some of the examples that is used for detecting these fraudulent are purchasing accounts payable, sales projections, warehouse movements, employee shift records, returns, and store-level video and audio recordings.

7.11. CONCLUSION

This chapter indicates the importance of customer acquisition and brings out certain dimensions that are relevant while acquiring a new customer in a retail store. It also explains how customer acquisition has become a difficult task for retails in today's world.

This chapter explains the importance of customer acquisition, different methods are used to acquire customer acquisition. As of today, retailers are fighting for the acquisition, as there is an increase in retails number and introduction of many brands. This chapter also explains the ways retailers are trying to acquire new age customers.

So, it concludes that customer acquisition has become a tricky task, as the new techniques are used to gather information about the customer requirement. Which help the retails to provide the product according to the customer need.

REFERENCES

1.   Customer Retention, (2011). [ebook] Emmanuel OSEI BOAKYE. Available at: https://www.researchgate.net/publication/285927801_Customer_Retention (accessed on 10 March 2020).

2.   Galetto, M., (2015). NGDATA | What Is Customer Acquisition? Definition and Process. [online] NGDATA. Available at: https://www.ngdata.com/what-is-customer-acquisition/ (accessed on 10 March 2020).

3.   Goel, A., (2019). How New-Age Analytics Can Transform the Retail Industry?[online] Indian Retailer. Available at: https://www.indianretailer.com/article/technology/back-end/how-new-age-analytics-can-transform-the-retail-industry.a6393/ (accessed on 10 March 2020).

4.   In.mathworks.com. (n.d.). What is Predictive Analytics?—3 Things You Need to Know. [online] Available at: https://in.mathworks.com/discovery/predictive-analytics.html (accessed on 10 March 2020).

5.   Juneja, P., (2015). Customer Acquisition—Meaning and Its Process. [online] Managementstudyguide.com. Available at: https://www.managementstudyguide.com/customer-acquisition.htm (accessed on 10 March 2020).

6.   Lonnie, K., (2018). What is Co-Creation and How are Companies Using it?. [online] KL Communications. Available at: https://www.klcommunications.com/what-is-co-creation/ (accessed on 10 March 2020).

7.   Marketsandmarkets.com. (n.d.). In Memory Analytics Market by Component&ApplicationGlobalForecast—2022|MarketsandMarkets. [online] Available at: https://www.marketsandmarkets.com/Market-Reports/in-memory-analytics-market-148739205.html (accessed on 10 March 2020).

8.   MicroStrategy, (n.d.). Cloud Analytics: Everything You Need to Know. [online] Available at: https://www.microstrategy.com/us/resources/introductory-guides/cloud-analytics-everything-you-need-to-know (accessed on 10 March 2020).

9.   researchgate.net. (2017). Retail Analytics: Driving Success in Retail Industry with Business Analytics. [online] Available at: https://www.researchgate.net/publication/323309092_Retail_Analytics_Driving_Success_in_Retail_Industry_with_Business_Analytics (accessed on

10 March 2020).

10. Retail Analytics: Game Changer for Customer Loyalty, (2014). [ebook] cognizant 20-20 insights. Available at: https://www.cognizant.com/ InsightsWhitepapers/Retail-Analytics-Game-Changer-for-Customer-Loyalty.pdf (accessed on 10 March 2020).

11. Satyanarayan, S., (2018). Here are 4 Retail Habits Being Adopted by the New Age Consumers. [online] Indian Retailer. Available at: https://www.indianretailer.com/article/multi-channel/eretail/Here-are-4-retail-habits-being-adopted-by-the-new-age-consumers.a5991/ (accessed on 10 March 2020).

12. Sisense, (n.d.). What is Retail Analytics? | Sisense. [online] Available at: https://www.sisense.com/glossary/retail-analytics/ (accessed on 10 March 2020).

13. Talisma, (n.d.). Rise of the New Age Customer – Era of Contextual Engagement. [online] Available at: https://www.talisma.com/rise-of-the-new-age-customer/ (accessed on 10 March 2020).

8

Future With Analytics

CONTENTS

In the chapter, the future with analytics, the analytics trends that will change the future is discussed. The chapter explains the expectations that businesses have from analytics and BI. The chapter also explains the future analytics environment in terms of analytics and data integration.

In addition, the chapter also explains why data quality is the biggest obstacle for companies. The chapter also highlights the need for business intelligence and analytics.Furthermore, the chapter also describes the models of disruption fueled by data and analytics. The chapter also throws some light on the role of analytics in changing digital marketing. In the end, the chapter explains some of the predictions of analytics that may occur in the near future.

8.1. INTRODUCTION

The potential of analytics has made a leap forward in recent years. In the past few years, with the increase in the volume of data, more complex algorithms have been developed, and computational storage and power and have been dramatically improved. These current trends triggering the advancement in technology and business disruption.It is important to note that companies as of today are capturing only a fraction of the potential value from data and analytics. The highest development has observed in location-based services and in retail, both areas with digital native competitors. In contrast, the public sector, manufacturing, as well as health care services have captured not captured more than 30% of the potential value that can be derived from analytics.The biggest barrier that the companies face in deriving value from analytics is in the form of integrating data-driven insights into day-to-day business processes.Another barrier is attracting and retaining the right talent—not only data scientists but business translators that can combine data-savvy in accordance with industry and functional expertise. In addition, some of the bigger companies are fully exploiting the use of analytics in improving their core operations as well as in launching entirely new business models.

Data and analytics are primarily known for disruption in businesses. Introducing new types of data sets ("orthogonal data") can disrupt industries, and massive data integration potential can breakthrough technological and organizational silos, enabling new models and insights. Hyperscale digital platforms can assess the needs of both buyers and sellers in real-time, and thus able to address the problems faced by both parties, thus transforming inefficient markets.

8.2. 8 ANALYTICS TRENDS THAT WILL SHAPE THE FUTURE

Figure 8.1. Future trends of analytics.

Source: Image by Pxfuel.

Analytics is the backbone of the future digital economy. In order to grab the opportunities in the future, it is important to recognize the growing trends that will have an overall impact on the analytics program.

Most business leaders all over the world are fully vigilant about the digital transformation that the businesses are going through, primarily because of the fact that it is the digital business that will decide the success of every company in the future.

As business leaders are concerned about how to drive their digital transformations, a top investment priority is in the sector of business intelligence (BI) and analytics.

Data is the fuel that gives direction to the digital economy and analytics is the process that transforms that raw material into usable and final assets, which is important for making key decisions. An in-depth understanding of the analytics trends is crucial for business leaders as well as their teams across all industries.

8.2.1. Analytics Is Becoming an Experience

Unlike BI hard work of the past (that basically focused on delivering attractive charts and graphs that envision the state of the business), analytics of the future will be based upon the contextual experience.

Nowadays, analytics is all about how information is received and used up as it is about the message to be delivered. A greater degree of personalization on the basis of context becomes a critical element of all the analytics program.

8.2.2. Location and Time of the User

Effective analytics is all about collecting the comprehensive details of the consumers that vary from the location of the consumer to the time of day in order to optimize the overall experience of users. For example, the time zone of the recipient is much essential as compared to the time zone of a report creator.

Location-aware analytics has also started gaining importance; the location of the users is now another input that regulates what information the user will receive and what not.

Figure 8.2. Use of analytics saves ample time for the users and professionals.

Source: Image by Pixabay.

8.2.3. Mobile Device Use Grows

As there is a significant increase in the digital devices used by consumers that range from mobile phones, glasses, smartwatches, digital personal assistants, in-car displays, and even video gaming systems, the future success of these devices relies on the impactful analytics.

Unlike the traditional era, when consumers mostly consume information on the computer that has a larger screen, nowadays they want all the information to be visible on the mobile screen and this can be possible only with the help of analytics. Information such as tables and charts are very difficult to show on a mobile screen as it is smaller in size but using analytics, it is possible.

Nowadays as the consumer is using multiple devices at the same time, it is important to provide personalized content not only in accordance with the user's characteristics but also with the user's device characteristics.

8.2.4. The Rise of the User Journey

The user journey has now relied more on the analytics. Recipients of analytics results include customers, internal business decision-makers as well as internal and external partners. For every business, these users are attached to them in one or another way.

All those historical interaction points create the journey that individuals have taken. With the evolution of analytics, data points from this journey will personalize what, when, and how information is delivered to the end-user.

Every business at some time need to interact with their customers that can be in the form of knowing their interaction history, past purchase behavior pattern, and what information has been sent to them creates a unique point-in-time need for information.

A B2B (business to business) relationship with a business partner or communication with an internal decision-maker must also take into consideration the user journey. A high degree of personalization will always be at the top in effectively delivering analytics to drive the growth in business.

8.2.5. Continuous Analytics

With the increase in the importance of the Internet of Things (IoT) in business and corresponding streaming data, the window to analyze, capture, and respond shortens. Analytics programs in the past were successful when they could bring the results in a few days or weeks, but the future will progress dramatically as these windows shorten to hours, minutes, and seconds—perhaps even milliseconds.

Figure 8.3. Data analytics provides path for continuous process of analytics.

Source: Image by tOrange.

The increase in the expectation of end-users to get the information faster and quickly put the pressure on analytics teams to find out how much analytical processing and refinement is enough.

It is vital to check whether simple processing that can be delivered the information in seconds is sufficiently effective as compared to the complete processing process that takes hours or even days to process and deliver the information.

There is also needed to check if the consumer can take some information now, that is relevant, and some later. The information that is simplest, requires most basic analytics in real-time and provide more comprehensive analysis later.

8.2.6. Augmented Data Preparation

During data preparation, machine learning (ML) automation is starting to augment data quality, and data profiling, enrichment, modeling, and metadata development and cataloging. Techniques that come into handy include supervised learning, reinforcement learning, and unsupervised learning and these are taking data preparation to the next level.

Unlike the processes used in the past, that basically rely upon rule-based approaches to make the data in presentation form, these advanced machine learning processes allows the data to become more adept at responding to changes in the data, especially outliers.

8.2.7. Augmented Data Discovery

In addition to allowing the users in the preparation of data, the advanced algorithms now enable information consumers to visualize and narrate relevant findings within the data in an easy and convenient way. These include enhancements such as automatically exposing correlations, clusters, exceptions, predictions, and links within the data without having end-users to write algorithms or build models themselves.

This augmented data discovery opens the door for the entry of citizen data scientists. These users consist of information consumers who, with the help of augmented assistance, will start to identify as well as respond to patterns in the data much more rapidly and in a much more distributed model than in the past (when an only limited number of data scientists are familiar with how to do this work).

8.2.8. Augmented Data Science

The advent of the citizen data scientist will not eradicate the requirement for a data scientist who investigates and research the data to find out profitable opportunities for business growth and development. As these dedicated data scientists have given the task of simpler work to citizen data, their examination becomes both more challenging as well as potentially more supreme to the business.

To improve their overall effectiveness, it is often seen that machine learning is applied more extensively in areas such as feature and model selection.

The application of machine learning in doing the repetitive task independently without the help of scientists allow these data scientist to focus on more important areas of their work: to identify unintuitive patterns in the data that is essential to take the business to higher level and then move these into an operational state in order to start earning profits.

8.3. THE EXPECTATIONS FROM DATA AND BUSINESS ANALYTICS

One thing that is certain and predictable in the future is that – data analytics will start gaining momentum in the next few years and will play a dominant role in providing new limitless technology solutions. The overall dependence of business on BI as well as Analytics will be increased and central to the success of a business.

But what will business analysis look like in the coming years? How the present scenario of BI and data analytics will evolve over a period of time, and how can companies make sure that the resource and technology they will use to remain competitive in the market will give the desired result? There's no doubt at the pace at which data analytics and BI tools are evolving.

> According to one report published by the McKinsey focused on the healthcare industry, "data analytics and information services will have the fastest growth rate at 16–18% over the next five years, while core administrative services are unlikely to see much growth due to automation."

Here are some of the key things that businesses should keep in mind in order to get success in the business.

8.3.1. Staffing Shortages Will Create Challenges

It is important to note that in the next few years, there will be a significant shortage of qualified data analysts and its effects are clearly observable in the market today. This may result in causing a huge impact on how the business works. To address this problem in the future, it is vital to start planning now, so that to become ready for the uncertainties.Planning can be in the form of developing a unique incentive program for the company's employees to remain competitiveness in the market, conducting training programs for employees to improve their skills, and motivate them to remain in the company for a longer period. The motivation can be in the form of monetary rewards, remuneration, recognition, promotion, and incentive.

8.3.2. Wider Adoption by Business Users

The importance of BI and analytics tools will persist only if it continuously focus on how to make it user friendly and increase the use of natural language that allows business users to extract data and create reports without getting into complex algorithms used in analytics.It will not only help in improving the efficiencies, but also ensures wider adoption by the companies, thus help in addressing some of the problems created by the data scientist shortage.

8.3.3. Increased Reliance on Large Data Networks

Vast data stores occasionally termed as advanced data networks, will become highly valuable for companies to access.

A large amount of data that can be accessed through analytics helps companies to evaluate and access the consumer needs and wants and bridge the gap if there is any, thus providing more personalized services and experience to their customers. It allows the companies to keep up to date about what is in trend in the market, so that they can deliver the basis on the basis of the same.

8.3.4. Growth in Machine Learning Will Accelerate

The opportunities provided by ML and artificial intelligence (AI) are limitless. It depends upon the companies how effectively they can grab these opportunities for their own benefits and thus creating new services that provide value in unique ways. Many business professionals are of the opinion that ML will take over the majority of customer service roles in the near future.

8.3.5. Managing Company Data Becomes Even More Challenging

Since the increase in the usage of data analytics and BI onslaught, it becomes prevalent for every company to manage source data and ensuring its consistence and accuracy. The validity of the data 'going in' determines the productive or useful data 'coming out.' As organizations rely heavily on the practicability of data to run their businesses, finding a way to solve this problem becomes non-negotiable.

With the increase in a variety of sources from where data come, it becomes difficult to identity to authenticity or originality of data, which further add challenge to the security and privacy issues of customers. Nowadays, as the customers become more conscious about their privacy, it becomes important to protect the same. Thus, analytics can be integrated into the system to take the data only from those sources that are verified.

8.3.6. Interconnectivity Becomes Critical to Success

With an increase in dependence on new internal tools for data analysis (DA) and BI, accompanied by an enhanced need to access data from external sources, networks, and IoT devices, interconnectivity will decide what the future of analytics will be as it plays an important role in building a cohesive data analytics machine for businesses.

To address the stiff competition that companies will face in the next few years, it is essential to formulate a plan to recruit skillful and talented people in the business as well as to plan well in advance for strategic investments. In addition, there is also needed to create process strategies for sustaining clean data across all systems.

8.4. THE FUTURE ANALYTICS ENVIRONMENT: ANALYTICS AND DATA INTEGRATION

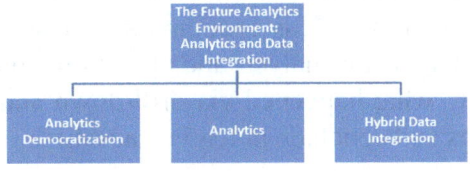

Figure 8.4. The future analytics environment.

There are a number of disruptive forces that are extensively affecting the ways how companies work, the way they design their organizations structure, develop, and deploy their analytics strategy.

During the past few years, data warehousing and decision support systems have become two critical elements of the corporate information intelligence arsenal, providing reporting and fundamental analytics processes in order to support their day to day operations.

Although the basic criteria of the companies' framework in which data sets are extracted from source systems, landed at a staging area for cleansing, standardization, and reorganization, and then encumbered into a monolithic data warehouse--has become a permanent standard, there are various disruptive forces in today's market that are driving changes in the organization functioning.

8.4.1. Analytics Democratization

Analytics democratization can be defined as the desire or need to provide reporting and analytics tools support to a broad range of citizen data analytics.

It is right to assume that companies will take into consideration different approaches for integration of these technologies, although there is a higher probability that all alternatives will eventually result in a hybrid environment that spans both on-premises and multiple cloud platforms.

While some companies will drift from their current on-premises reporting applications to the cloud (also known by the name to as "lift and shift"), while others will likely reconsider their analytics and reporting needs and revolutionize their environment to address both anticipated and ongoing future analytics needs. In this section, the primary focus will be on two facets of a future analytics environment that is arrayed on the cloud: analytics and data integration.

8.4.2. Analytics

Most, if not all, conventional analytics and data warehousing platforms support descriptive analytics. This encompasses most of the past adopted techniques such as diagnostic analysis and operational reporting that allows their clients to study what happened via drill-down and discovery, and to examine correlations and conclude potential causality.

The future scenario should be advanced to incorporate a full spectrum of analytics capabilities and applications, including:

- **Predictive analytics**, which uses ML, data mining, and AI techniques to develop strategies and models for assessing future behaviors.

- **Prescriptive analytics**, which is basically concerned with providing optimal outcomes of particular alternatives based on predictive analytics. In other ways, it can be said that prescriptive analytics plays a dominant role in automating decision processes.

- **Integrated analytics**, which is basically interested in integrating analytical models within information flow to ensure the support of automated decisions in execution.

- **Feature extraction and text analytics**, which play a very bigger role in identifying and extracting features from unstructured or semi-structured data that can then be used to trigger prescriptive and predictive analysis.

Figure 8.5. Various capabilities in the field of analytics.

While previously used data warehouse architectures could be lifted and shifted to the cloud, some of the updated data management architectures (such virtualized access to object storage and as in-memory hybrid databases) that are explicitly created to exploit the use of scalable cloud resources in order to integrate streaming data with AI, and ML models and algorithms.

8.4.3. Hybrid Data Integration

It is important to note that there are two elements of "hybridization" in the evolving extended information environment. First, the enterprise is growing at a rapid rate beyond what is said traditional on-premises configuration.

This rapid growth directs them to migrate their data and application to one or more cloud platforms, which triggering to adopt a more complex hybridized computing environment.

Second, analytics applications and information management are increasingly capable of allocating and processing structured data, as well as semi structured data (for instance JSON or XML documents) and unstructured data assets (such as transcriptions or freeform text of audio data).

This necessitates a greater need for enterprise-wide data awareness. With the increase in the size of the analyst community, it becomes essential on the part of each individual to rapidly determine which data assets are easily accessible for use in addition with the appropriate metadata that direct the data consumer in the use of data asset.

And, with the increase in the sources of real time streaming data, analysts are in need of tools to help quickly gather and analyze these data feeds. The future analytics environment must be capable enough to accommodate integration across these two dimensions, incorporating strength and abilities such as

- **Data discovery** plays a very important role in allowing analysts to get an insight into the contents of collected data assets and help to collect and characterize structural metadata as well as evaluating whether the data asset holds delicate information, which requires privacy and security.

- **Virtualized data accessibility** minimizes the effort of the performer to, again and again, copying data from one platform to another by letting analytics applications to access data in place.

- **Data pipeline orchestration tools** are primarily responsible for managing the increased complexity of simultaneous ingestion/processing/analysis applied to both streamings as well as static data.

- **Data catalogs** are basically concerned with providing a repository for data asset metadata that from time to time signals consumer communities about available data assets and how it can be used in order to get the desired result.

- **Real-time data ingestion** makes the task of the data analysts easy in terms of developing applications that integrate batch data with continuously streaming data on a real time basis.

8.5. DATA QUALITY: COMPANY'S BIGGEST OBSTA-CLE

Data analytics is becoming increasingly sophisticated, scalable, and accessible aided by cloud infrastructure, but still there are various challenges exists in data quality as well as building the right culture

The market of data analytics has evolved rapidly, making use of algorithms to larger datasets on a more frequent basis by data scientists as well as the number of people associated with this.

Many of the sophisticated algorithms used in the analytics sector are now more than 50 years old, but the recent advancement in the storing system to store unlimited data and enhanced compute power, and improved development tools, has enabled greater scale today and for the future of data analytics.

Data analytics have improved significantly to apply more complex analyses in a much more responsive manner. Analytics is all concerned with creating insight and giving solutions to critical problems, which can be exploited by businesses to achieve the desired result.

Contemporary systems can now integrate machine-learning into more diverse datasets, where the analysis has relied more on the computer instead of the operator. The real-time arrival of new data enables insights to be discovered as soon as they occur.

8.5.1. Companies Looking to Make Better Use of Data

Historically, the overall needs and demands of corporate data have been less onerous. But as now companies have started to leverage their data, rather than just collecting it, one of the most prevalent issues is the veracity of that data.

The proper analysis of the recorded data is as important as the quality of the data fed into a system. Improving the overall quality and effective management of data has not yet considered as a requirement, but this leads to the formation and application of data that isn't up to scratch for the type of analytics organizations need to develop.

Analytics tools and technologies are to a larger degree commoditized, but the critical challenge for the companies to exploit them is the formation and availability of adequate datasets. Establishing these foundations is usually a specific problem in businesses that have not historically fostered a data-centric culture.

"Sometimes relevant data doesn't exist and when it does its location is often poorly understood, riddled with quality issues and spread across multiple systems of record," says Paul Fermor, UK solutions director at Software AG. "The majority of organizations that have realized the potential value of their data are engaged in substantive projects to improve its quality, real-time availability and integration across systems."

8.5.2. Is Cloud the Future of Data Analytics?

It is generally seen that business leaders want data analytics potential to get benefits in their organization, however, they often got irritated with slow progress because of the underlying shortcomings of their existing core data infrastructure. This practice is mostly seen in companies that have grown through mergers and acquisitions, leading to fragmented technology, cultures, and teams.

The rise of the cloud has helped resolve data infrastructure scalability concerns, providing data analytics software as a service. Cloud has ensured the latest tooling is readily available without the need to maintain and patch, while traditional database administrators can build machine-learning models without the knowledge required just a few years ago.

"Cottage industries and data fiefdoms will gradually disintegrate; the future of data analytics is in the cloud," says James Tromans, technical director at Google Cloud. "Those with the correct clearance can quickly start applying advanced data analytics to a valuable business problem in a way that simply wasn't possible previously."

Sometimes there is a very low possibility of getting the relevant data and when it does, its location is often poorly understood, having certain quality issues

The integration of technological advancement into companies allows them to grow and become competitive with the help of cloud infrastructure. Increased availability of the infrastructure that algorithms reside in means chief cloud services providers are heading towards using related technology such as live streaming, security or privacy issues. As well as the analytics, they are interested in covering the control and regulatory compliance of the data.

8.5.3. Smart Data Analysis Key to Beating Competition

An example of a company that already started combining ML and data analytics is Ocado Technology. In collaboration with Big Query, Ocado developed a mechanism for foreseeing and identifying fraud cases among millions of other normal events and evaluating and assessing the past order history of customers such as share of card details, account information, etc.

By creating advanced models and techniques, Ocado improved its overall efficiency and effectiveness of detecting fraud by a factor of 15.

Businesses and industries are continuously facing tough competition in the form of disruption with the increase in the technology complexities. How companies can effectively address these challenges is dependent on their ability to manage and exploit their data says Nick Whitfeld, head of data and analytics at KPMG UK.

He is of the opinion that the culture around data needs to change. There is a need on the part of organizations to understand whether the quality of data being generated in not appropriated, this will undermine investment or focus in the future of data analytics.

Mr. Fermor at Software AG believes the future of data analytics will address some of the bigger problems, such as creating more human-like machines. "This might manifest itself in more convincing chat bots and AI assistants or improved medical diagnosis tools.

There are also efforts to automate the machine-learning process, which is still driven by humans, and create a less technical, self-service approach to creating and deploying sophisticated models," he says.

Organizations that are seeing their future in data analytics should need to develop their skills in the following sector and develop operating models in the future of data analytics.

Ensuring that companies can successfully process, and exploit data quickly will require an environment of talented human resources, who are self-motivated to work at an extraordinary level of accuracy.

8.6. THE NEED OF BUSINESS INTELLIGENCE AND ANALYTICS

The term intelligence, BI has started gaining popularity as early as the 1990s. Recently to describe the data sets, file formats, data schema, and analytical tools in applications that are so complex and large (from charged coupled devices, multimedia to electronic media data) that they require advanced and independent data storage, data process, management, and visualization technologies, BI and analytics have been used.

8.6.1. Applications of Business Intelligence and Data Analytics

A few multinational companies and IT patterns have helped shape over a wide span of time BI and Analytics look into bearings. Worldwide inventory network, rapid system relations, worldwide travel, and Customer relationship the board, outsider contracts have made an enormous open door for IT progression.

In the present scenario, the Data science period has inconspicuously slid on numerous networks, from an open segment, private area, and e-business to wellbeing part. Some of the high-sway and promising BI domains are discussed below:

- E-Business and Market Intelligence;
- e-Governance;
- Research in science and technology; and
- Public health and the environment.

8.6.1.1. Improving E-Commerce and Market Intelligence

It is often seen that the most primary use of BI and Data Analytics is at the e-commerce platforms and Web forums. For instance, E-commerce website leaders such as Amazon or eBay have developed their niche with the help of BI and Data Analytics.

In addition, it is also seen that the online streaming platforms such as amazon prime video, Netflix are also using analytics to present those contents to their customers, which are preferred by them.

The need for customer-generated ERP tool content on several platforms, social media platforms, chat groups, and crowd-sourcing systems presents another opportunity for practitioners and researchers to listen to the voice of the market from a variety of business constituents that includes employees, customers, investors, and the media.

Product Recommended systems have been designed by the researchers based on analytical techniques, such as database segmentation, association rule mining, supervised, and unsupervised learning techniques, support vector machines, anomaly detection, and graph mining.

8.6.1.2. Improving E-Governance

The advent of new Enterprise Resource planning tools opens the doors for reinventing good governance. There are some countries that have effectively become successful in ensuring E-Governance in their organization, while others that are still not making use of BI and Analytics to integrate e-governance into their system

In addition, there is wider scope for adopting BI and analytics research in e-governance policies, and politics, and applications since the government and the political processes have become more transparent, participatory, online, and multimedia-rich. Selected social media analytics (SMA) techniques, opinion mining, and social network analysis allow to better serve their target groups.

8.6.1.3. Improving Research in Science and Technology

Science and technology are the driving forces for current socio, economic, and cultural changes in society. Science and technology of today's era are being transformed by the new avenues through the use of enormous, heterogeneous data.

Various fields of specializations in Science and technology are gaining the advantages of high throughput sensors and instruments, from nanoelectronics, astrophysics, and climatology, to ecology and genome research.

8.6.1.4. Improving Healthcare and Public Health

The application of big data and analytics in improving rural health, wellbeing, and livelihoods by assessing the historical data and how to address the problems that may occur in the future.

It can also keep a check on a hike in real healthcare costs, so that to design subsidies or relief program for the poorer section of society. It looks at how more appropriate and efficient use of big data will facilitate to solve these and other health issues.

8.7. MODELS OF DISRUPTION FUELED BY DATA AND ANALYTICS

Data and analytics are already transforming the number of industries and these effects will be quadrupled in the coming few years, extending into new sectors and in many parts of society. This section considers how these changes are carried out.

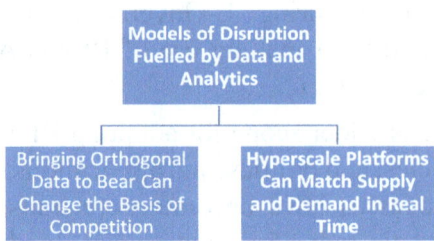

Figure 8.6. Models of disruption.

The focus of this section will be on the disruptive models that data and analytics enable rather than assessing its effects on each industry. Certain unique characteristics of the data and analytics market open the door to data-driven disruption.

For instance, it is often seen that data collected for one purpose can be deployed, for other unrelated purposes, in a completely different industry.

Using Google search data to form a price index and collecting credit scores of individuals to inform auto insurance rates are some of the paramount examples of how "orthogonal data" can be put to work in solving various types of business problems. Collecting a variety of data could set off cascading industry effects.

In some domains, supply, and demand are often mismatched, which results in not fully utilized the available valuable resources. Digital platforms that provide the services of largescale, real-time matching with dynamic pricing could transform markets with these characteristics—a list that covers various sectors such as hospitality, transportation, energy, certain labor markets, and even some public infrastructure.

In markets where officials have primarily dependent on blunt measures or broad demographics to categorize their customers into different segments, new players could completely disrupt the market by using finer behavioral data to customize products, services, as well as marketing activities for individuals.

It may result in causing noteworthy impacts in many areas, including media, health care, consumer products and retail, and political campaigns and even the education sector. In industries, where making use of data and analytics can significantly improve the performance—such as insurance, banking, retail, the public sector, and beyond—the organizations that are one step ahead in this can get the maximum advantage.

Data and analytics could have an impact majorly on those industries where the business model is built on innovation and discovery. ML and deep learning could trigger anything from drug development to engineering to customers' experience and design.

In addition, any attempt that can be marred by human error, fallibility or biases could be alleviated by one of the most profound capabilities of data and analytics: the ability to support, enhance, and even systematize human decision making by relying heavily on vast amounts of data and evidence.

Nowadays, technology provides vast opportunities in a way by addressing the number of human limitations from many situations and to make more accurate, faster, consistent, and transparent decisions. This capability has wide applications in every sector of businesses as well as in various aspects of life and society.

8.7.1. Bringing Orthogonal Data to Bear Can Change the Basis of Competition

Data and analytics have been part of the operating model to some extent in various industries over many years.

Insurers were offering services in the segment of the property, automobile, and casualty insurance, for instance, have long been assessing and incorporating risk factors such as demographics (place of residence, age, total driving experience, and so forth) into the analytics that direct theirs overall decision-making process. But as it is noted above, data are proliferating.Many new types of data, even from the new sources, can be used to address or effectively solve any problem. In industries where most business leaders are dependent on standardized or authentic data to take certain key decisions, bringing in new types of data sets in exchange for those already in use can change the basis of competition.

New players entering into the market having access to all these types of "orthogonal" data may pose serious threats to the already existed player on the market.

New companies that entered the marketplace with telematics data provide insight into driving behavior. This data set is orthogonal to the demographic data that had been used previously for underwriting. Other domains could be productive in getting orthogonal data from the internet of things (IoT).

Health care had previously dependent on medical histories, laboratory results, and medical examinations, but a new set of orthogonal data now being produced with the help of digital health devices used by consumers such as wearable's and connected health devices in the home (such as insulin pumps, ECG devices, and blood pressure monitors).

Some experts are investigating whether the data collected from these devices, while not clinical grade, could enhance wellness and overall health. Connected light fixtures, which identify the number of people being present in the room and have been sold with the promise of reducing energy usage, generate "data exhaust" that property managers can use to optimize physical space planning in future real estate developments.

Even in some companies, it was observed that human resource department made it mandatory for their employees to wear devices that capture data and yield insights into the "real" social networks that exist in the workplace, allowing these companies to improve interaction among their employees through necessary changes in work spaces.

Orthogonal data will rarely substitute the data which is already in use in a domain; there is more probability that an organization will integrate the orthogonal data with their current data

8.7.2. Hyper Scale Platforms Can Match Supply and Demand in Real Time

Data and analytics are completely transforming the way how markets connect sellers and buyers for many products and services. In some markets, as it is difficult to predict the behavior off sellers and buyers due to various factors, analytics help to address many of these problems by analyzing past historical trends.

It is often seen in markets, where there is the limited number of sellers for a product and the product has no repetitive buyers, which led to poor understanding between the buyer and seller. This is the case in real estate, for example, where buyers have strong preferences and finding exactly the right house is the priority.

In others, the speed of the match is critical. In this, the "Hyperscale" digital platforms come into effect by using data and analytics to meet both types of needs. These platforms have already addressed the majority of problems prevalent in urban transportation, retail, and other areas for many years.

But that could be only the beginning. In the future, they will also be used in transforming energy markets by enabling smart grids to supply equal energy to many small producers. And also, they can make the labor markets more efficient, altering the way how workers and employers connect for both traditional jobs and independent work.

8.8. THE ROLE OF ANALYTICS IN CHANGING DIGITAL MARKETING

Data science and ML are gradually changing the way how the industry works. From fashion to finance, technological capabilities are permitting users by allowing them to accomplish more in limited time.

Before the advancement in technology, data scientists spent almost 80% of their time doing preparation work. Now, calculations that take weeks, or even in some cases months, can be done instantly with the help of analytics.

The ability for business owners to introspect, comprehend, and apply their analytics for the growth and development of their brand has never been an easier task, and that's why the future of the analytics industry is so promising.

With the help of data analytics tools, a business leader can make more accurate decisions with proven results about their digital strategy that will provide them a competitive edge over their competitors.

Analytics has always and will always contribute heavily in determining the success of digital marketing because it allows companies to create a data-driven marketing strategy.

As marketers should be well aware of how much they are spending in marketing activities and what is the expected return, it is essential to keep track of all these activities and it can be possible only with the help of analytics. Whether it's likes, shares or conversions, the future of the analytics industry with data science and AI is changing the way how marketers perform their jobs.

8.8.1. Diagnostic Analytics

Diagnostic analytics is relevant only when there is adequate historical data available to fully comprehend why something has happened. For example, if a company-sponsored a social program and ran an ad campaign with a country setting in the background, which results in increased sales in suburban areas over urban areas, diagnostic analytics will help in identifying the reason for the same.

It allows the company to identify why the campaign attracts the suburban customers more than urban customers. This diagnostic process is important as it allows the company to start another campaign aimed at a particular audience that was unaffected by the previous campaign.

8.8.2. Descriptive Analytics

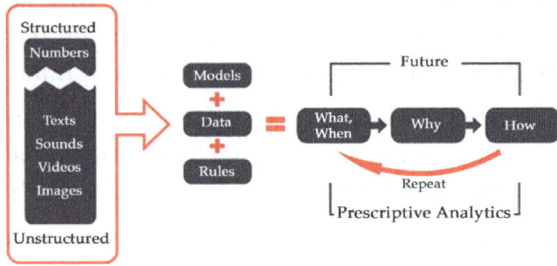

Figure 8.7. Prescriptive analysis process.

Source: Image by Wikimedia Commons.

Descriptive analytics refers to the data that is collected the during the preliminary stages of data processing and collection. It's raw data that can help in taking the basic decisions such as where your users are going and what they're doing there and elsewhere on the internet. With relevant time and information, descriptive analytics allows taking more informed decisions about the future of data analytics for business.

8.8.3. Predictive Analytics

Predictive analytics, as one can guess from its name, is related to prediction. The data under these analytics shows what most likely to happen in the near future based on descriptive and diagnostic analytics. It allows business leaders to assess the expected sales in different regions, by segmenting the audience on the basis of various factors.

With predictive analytics, one can predict the expected return on investment (ROI), so that to decide is it feasible to invest in a particular project. This helps in taking the decision in advance and alter things, so that to prevent any loss in the future.

8.8.4. Prescriptive Analytics

Prescriptive analytics as suggested by its name refers to prescribing the problem. How can a company work on their sensitive areas based on the provided data about the sales cycle? This is where augmented analytics and ML come into effect; these technologies can provide direction about the next possible move.

Based on the data processing, ML and AI allow the organization to spot sore areas. The future of business analytics (BA) relies on the prescriptive analytics because it is the last stage where crucial strategies and policies are formed.

8.8.5. Artificial Intelligence and Machine Learning

Augmented analytics refers to the solution that is technology-driven which transforms business ideas into business capabilities. Augmented analytics is the ability to evaluate, analyze, assess, review as well as interpret the data with the help of ML and AI in order to support business leaders to do actionable steps, which enhance their business growth and development.

The real magic of these technological tools is clearly observable because these analytics tools operate within the context of business trends that can be accessed from the brand's data.

These advance tools not only help in saving time and money, but allows to spend time on more important issues, and thus enhancing the profits of the company. These analytics are beneficial to every type of business whether it is small scale or large scale, or any nature of workings.

Corporations and entrepreneurs alike can restructure their workflow by permitting AI to do all the heavy lifting while giving them enough time to work on executing the plans.

Of course, every business should review various options before opting for the one and select that is more appropriate for the business. It should be reliable for the current as well as for the future scenario. But apart from some review work, there is not so much risk associated with choosing an option as the data is available.

This changes the game for digital marketers as it allows them to spend less time hashing out the particulars of their next campaign and devoting more time to the execution of the next advertisement they're going to run.

8.9. 6 PREDICTIONS OF ANALYTICS

Analytics have become an integral element for almost every company to engage in the decision-making process. The importance of analytics has grown in the last few years and now companies to get a competitive edge adopting it on a wider scale, especially those companies that use data assets as a core competency and point of origin. Six predicted trends of the analytics that are expected in the near future are explained below:

8.9.1. Deliberate Data-Culture Initiatives

Companies nowadays are recognizing the need to shift their organizations' culture to become more data-inspired in decision-making with respect to core decisions such as tactical and strategic decisions.

Companies, as facing stiff competition, have started taking the initiative to integrate data culture in their organization. some companies that have experienced employees in n this field have already adopted data culture in their organization, which others are either outsourcing these services or engaging in the training sessions to enhance the skills of their employees to increase absorption and adoption of analytics solutions.

8.9.2. Unstructured Data Proliferation

With the dynamic shifts in the method of data collection, types, as well as availability of data on consumers, machines, and just about everything else, it is right to assume that the future is more vibrant with respect to holding more creative types of data, especially unstructured elements such as video, audio, mood capture, and other relevant data.

8.9.3. Need for Real-Time Models

Engaging in planning to develop a solution for a problem that may occur in the future would be helpful but getting prepared in advance by evaluating the current factors is more helpful to address the problem in real time. This trend across analytics will continue in the future. For instance, consumers may have higher expectations from a brand in terms of after sale services, durability, and recognition to fulfill preferences at the moment of relevancy.

Analytics on demand and in real time may take over traditional static insights to meet the fast pace environments where they are applied, whether that's a market environment, research lab, or other.

8.9.4. Specificity, Granularity of Insights, including Mass Personalization

The precision and granularity of analytics solutions will likely progress in the coming years as the focus will shift on more details and complex data on individual units, whether that includes rail cars, drops of a poisonous substance, or persons. Analytics, as it is widely known, provides unique solutions on a per-unit basis to create more precise reactions.

8.9.5. Tool Reliance/Citizen Analyst

Accepting secret sauce as a principal element may become more common, and confidence in packaged analytics processes could breed more citizen analysts across businesses. This trend might be challenging for some businesses, but the increasing demand for analytics solutions and work may result in more open-mindedness to apply sophisticated tools in the hands of professionals with less data science knowledge.

8.9.6. Increased Movement toward Automation, AI, Deep Learning Techniques, Methods, and Processes

As companies are, in the current scenario, heading towards developing specific, real time solutions to answer more critical and complex problems, there is a need to apply, or even discover more sophisticated methods to answer or find a solution to these questions.

The improved capabilities and dynamic nature of analytics continue to encourage and allow more companies and even individuals to do their work in an efficient and effective way. The future allows business leaders to focus their time on more important areas by enabling analytics to perform some of the major tasks independently.

8.10. CONCLUSION

In the end, it is concluded that analytic has a vibrant future, and its role will continue to increase in the future. There are many companies that have already adopted analytics in their business and have extracted the benefits from the same. With the increase in competition, it is more likely that more companies will gradually integrate analytics and BI into their business.

As analytics is evolving, its role in every sector, whether it is business, education or healthcare department, will continue to increase. It is not easy to maximize the use of analytics unless there are experienced professionals to integrate the analytics and know its applications. Therefore, those who want to grab the opportunity that analytics will provide need to work in advance for the same.

Analytics plays an important role in doing various tasks of businesses automatically, that otherwise required huge efforts, therefore the investment that the company will be incurred in implementing analytics in their business will pay off in the future. Thus, in the long run, analytics will always provide benefits and as in the future, it plays a more dominant role, its application in the business is essential for survival.

REFERENCES

1. Analytics, (2020). The Future of Data Analytics – Compact. [online] Compact. Available at: https://www.compact.nl/articles/the-future-of-data-analytics/ (accessed on 10 March 2020).

2. Analytics and complexity: Learning and leading for the future, (2020). [ebook] Available at: http://www.ascilite.org/conferences/ Wellington12/2012/images/custom/beer,colin_-_analytics_and.pdf (accessed on 10 March 2020).

3. KDnuggets, (2020). The Future of Analytics and Data Science— KDnuggets. [online] Available at: https://www.kdnuggets. com/2019/09/future-analytics-data-science.html (accessed on 10 March 2020).

4. Keller, J., (2020). Future of Business Analytics Trends in 2020 and Beyond. [online] Selecthub.com. Available at: https://www.selecthub. com/business-analytics/business-analytics-trends/ (accessed on 10 March 2020).

5. Knight, M., (2020). The Future of Analytics: What is All the Hype About? – DATAVERSITY. [online] DATAVERSITY. Available at: https://www.dataversity.net/future-analytics-hype-real/# (accessed on 10 March 2020).

6. Loshin, D., (2020). The Future Analytics Environment: Analytics and Data Integration. [online] IT Pro. Available at: https://www. itprotoday.com/data-analytics-and-data-management/future-analytics-environment-analytics-and-data-integration (accessed on 10 March 2020).

7. Maguire, J., (2020). 5 Data Analytics Trends Shaping the Future of Analytics. [online] Datamation.com. Available at: https://www. datamation.com/big-data/5-data-analytics-trends-shaping-the-future-of-analytics.html (accessed on 10 March 2020).

8. ResearchGate, (2020). The History, Evolution, and Future of Big Data & Analytics: A Bibliometric Analysis of its Relationship to Performance in Organizations. [online] Available at: https://www.researchgate. net/publication/329118026_The_history_evolution_and_future_of_ big_data_analytics_a_bibliometric_analysis_of_its_relationship_to_ performance_in_organizations (accessed on 10 March 2020).

9. Romeike, F., & Eicher, A., (2020). Predictive Analytics: Looking into the Future. [ebook] Available at: https://www.risknet.de/fileadmin/

user_upload/Elibrary/Predictive-Analytics_Romeike_FIRM-Jahrbuch-2016_ENG.pdf (accessed on 10 March 2020).

10. Rossi, B., (2020). Is Cloud the Future of Data Analytics?. [online] Raconteur. Available at: https://www.raconteur.net/technology/future-data-analytics (accessed on 10 March 2020).

11. Silva, T., (2020). The Future of Business and Data Analytics—2019 | ClicData. [online] ClicData. Available at: https://www.clicdata.com/blog/the-future-of-data-and-business-analytics-whats-coming-and-what-you-need-to-know-now/ (accessed on 10 March 2020).

12. Soni, A., (2020). 10 Trends that Would Shape the Future of Data Analytics. [online] YourStory.com. Available at: https://yourstory.com/2017/12/data-analytics-future-trends (accessed on 10 March 2020).

13. The Age of Analytics: Competing in a Data-Driven World, (2016). [ebook] McKinsey & Company 2016. Available at: https://www.mckinsey.com/~/media/McKinsey/Business%20Functions/McKinsey%20Analytics/Our%20Insights/The%20age%20of%20analytics%20Competing%20in%20a%20data%20driven%20world/MGI-The-Age-of-Analytics-Full-report.ashx (accessed on 10 March 2020).

14. Transforming Data with Intelligence, (2020). 8 Analytics Trends That Will Shape Your Future | Transforming Data with Intelligence. [online] Available at: https://tdwi.org/articles/2019/01/02/adv-all-8-analytics-trends-that-will-shape-your-future.aspx (accessed on 10 March 2020).

15. Wright, B., & Wright, V., (2020). The Future of Analytics: 6 Predictions – SD Times. [online] SD Times. Available at: https://sdtimes.com/data/the-future-of-analytics-6-predictions/ (accessed on 10 March 2020).

16. Edwards, J., (2020). Predictive Analytics: Transforming Data into Future Insights. [online] CIO. Available at: https://www.cio.com/article/3273114/what-is-predictive-analytics-transforming-data-into-future-insights.html (accessed on 10 March 2020).

17. Sisense, (2020). Augmented Analytics: The Future of Business Intelligence l Sisense. [online] Available at: https://www.sisense.com/whitepapers/augmented-analytics-the-future-of-business-intelligence/ (accessed on 10 March 2020).

18. Soni, A., (2020). 10 Trends that Would Shape the Future of Data Analytics. [online] YourStory.com. Available at: https://yourstory.com/2017/12/data-analytics-future-trends (accessed on 10 March 2020).

INDEX

Printed in the United States
By Bookmasters